The Liberty Hyde Bailey
Gardener's Companion

The Liberty Hyde Bailey Gardener's Companion

Essential Writings

Edited by John A. Stempien
and John Linstrom

Comstock Publishing Associates

an imprint of

Cornell University Press

Ithaca and London

First published 2019 by Cornell University Press

Printed in the United States of America

Library of Congress Cataloging-in-Publication Data

Names: Bailey, L. H. (Liberty Hyde), 1858–1954, author. | Stempien, John A., 1967– editor. | Linstrom, John, 1987– editor.
Title: The Liberty Hyde Bailey gardener's companion : essential writings / edited by John A. Stempien and John Linstrom.
Description: Ithaca [New York] : Comstock Publishing Associates, an imprint of Cornell University Press, 2019. | Includes bibliographical references and index.
Identifiers: LCCN 2019008042 (print) | LCCN 2019009152 (ebook) | ISBN 9781501740275 (pdf) | ISBN 9781501740282 (epub/mobi) | ISBN 9781501740237 | ISBN 9781501740237 (cloth ; alk. paper)
Subjects: LCSH: Bailey, L. H. (Liberty Hyde), 1858–1954. | Gardening. | Gardening—Literary collections. | Gardens. | Gardens—Literary collections.
Classification: LCC SB455.3 (ebook) | LCC SB455.3 .B35 2019 (print) | DDC 635—dc23
LC record available at https://lccn.loc.gov/2019008042

*This collection is dedicated
to the spirit and legacy
of Liberty Hyde Bailey
and to amateur gardeners everywhere.*

The amateur is the ultimate conservator of
horticulture.

Liberty Hyde Bailey, *The Garden Lover*, 1928

It is hoped the book will contribute to the understanding and the dignity of plant-growing. The grower should be proud to be in the company of so many kinds of plants.

Liberty Hyde Bailey, *Hortus Second*, 1941

Contents

II. The Growing of the Plants

III. Flowers

IV. Fruits & Vegetables

V. Spring to Winter

VI. Epilogue

Illustrations

Full publication details for the figures listed here can be found in appendix 2.

Preface

Little children love the dandelions; why may not we?
Love the things nearest at hand: and love intensely.
Liberty Hyde Bailey, *Garden-Making*

December 1861 was a dark month in the lives of one family on the western frontier. Much like the snows of the Michigan winter that had begun to settle around their small, boxlike frame house, erected just four years earlier to replace the drafty log cabin that had preceded it, a grave sickness began to settle into the family. The deep rashes and strep throat of scarlet fever passed from one child to the next, infecting all three sons and leaving only the anxious mother and father unafflicted to care for their children. As winter wore on, the eldest son's health continued to deteriorate, and in February the fourteen-year-old died at home, in the old custom, surrounded by his grieving parents and his two younger brothers, one thirteen years old and the other only three.

That three-year-old child would not remember that day, but he would remember the tragedy that struck in the following December, when he stood with his father and brother beside the sickbed of his mother, Sarah, and watched her "pass into the silence," the victim of diphtheria, a disease without a cure. A woman of artistic sensibility, who had learned to write as an adult and even took to composing poetry to share with her family, Sarah never recovered from the grief of losing her eldest son, but she left behind traces of the artful grace with which she had cared for the frontier farm home. Among those traces was a small garden of hardy perennial flowers in front of the house, facing the path that led into town. When planting that garden, Sarah had let her beloved youngest son drop the seeds into their places, and now that she was gone, her surviving husband took time out of his regular farm work to help the four-year-old care for that garden—especially his favorite flowers, the bright splashes of the little "grass pinks" (*Dianthus plumarius*, more commonly "cottage pinks"). So life grew in the wake of loss, and the young child, known by the family as little Tom, by the children of the neighborhood as Libby, and on his birth certificate as Liberty Hyde Bailey, learned to love the rhythms of the garden.

I was fresh out of college in the bright summer of 2010 when my mother and I found ourselves standing in front of that same garden plot, listening to John Stempien, now coeditor with me of this volume, tell that story to us. While perhaps no one in my family can claim the kind of green thumb that seems to have run through the Bailey family, my own childhood memories began to well up— of my mother's annual outdoor pots of basil and the fresh pesto that would become my favorite childhood food, and of sitting with my father in the side garden with a spade and too-big gloves, trying to figure out which plants were weeds and digging holes just the right depth (or so Dad affirmed) for daffodil and daylily bulbs. The three of us, Mom, John, and I, stood there for a moment, silently gazing at the historic garden bed where members of the Liberty Hyde Bailey Museum had recently replanted a garden of "grass pinks" in honor of Sarah Bailey's memory.

John, who had been director of the birthplace/childhood-home museum since 2006, as well as its unofficial docent, broke the silence. "That's really the beginning," he said. "This museum, and Bailey's whole lifework, sort of starts right here."

"Dianthus plumarius *in the wild*."

Despite having moved to South Haven, Michigan, when I was four, Mom and I hadn't really been tuned in to the little museum until that spring, when my father read something in the paper about a book talk being given by John Stempien at the local community college. John had just two years earlier facilitated the reprint, by Michigan State University Press, of Bailey's manifesto of environmental philosophy, a slim little hundred-page volume titled *The Holy Earth*, which had been effectively out of print since the 1980s. This piqued our interest—Mom, an elementary school teacher at South Haven Public Schools, wanted to develop an outdoor learning center to use with her kids, and since I was about to head out to Iowa for an MFA program in Creative Writing and Environment, I figured I might as well get to know something about this seemingly obscure environmental writer from my hometown. Like Mom, John also taught in the public schools (the museum being open only in the summers), in the nearby town of Lowell, and he loved the idea of a Bailey-inspired outdoor learning center at North Shore Elementary in South Haven.

"For Bailey, the love of nature always began with the child," he told us excitedly, and he went on to describe Bailey's role as a popularizer of the nature study movement at the turn of the last century, founding the influential nature study program at Cornell University that was led by the legendary Anna Botsford Comstock and "Uncle John" Spencer. Bailey's 1903 book *The Nature-Study Idea* contains some of his most delightful writing, as he recounts the many experiences he had working with both children and teachers, as well as his own childhood education, wandering the woods and fields surrounding South Haven and working as a farmhand and increasingly expert apple grafter on his father's farm.

John led us around the old homestead to where an ancient smokehouse of bright red handmade bricks still leaned under a

massive black walnut tree that the Baileys had planted a century
or more ago. The inside felt cool as we poked our heads in, and it
smelled like old leaves and mouse droppings. With the busy M-43
highway buzzing behind us, leading from South Haven to Kalamazoo (where Bailey's mother was born and his father started his
first orchard) and then northeast to Lansing (where Bailey would
eventually leave the farm to attend the State Agricultural College,
now Michigan State University), and the modern white-and-gray
facade of the town's hospital standing next door where South Haven's first commercial apple orchard once grew, we tried to imagine
an 1850s farmstead that operated before refrigeration, where any
preserved meat had to be cured in this small brick building the size
of an outhouse, before electricity or running water, surrounded by
miles and miles of forest. Pointing to the northeast, John described
the Potawatomi encampment that existed during Bailey's childhood. The local Potawatomi tapped maple trees in these woods,
maybe even the ancient sugar maple leaning over the house's
north side, and as a child Bailey would join the Potawatomi kids
in catching passenger pigeons, back in the days when their yearly
migrations would darken the sky for days. "I knew the Indians," he
would later affirm, "and I picked up something of their outlooks."[1]

John then led us into the old farmhouse that had been slowly
converted, by generations of caretakers, into a museum devoted to
Bailey's life and work, and that in 1983 was placed on the National
Register of Historic Places. Moving among various tables and
display cases, he pointed to old photographs and lifted artifacts,
telling the story of Bailey's life championing the cause of rural
America—which then represented the majority of Americans
and, often, the least privileged. For Bailey, the popular country life
movement of the early twentieth century was about more than just
providing rural access to urban conveniences—it was that, but it

was more importantly about preserving and celebrating the core of American culture, or what often today gets termed "folk culture." And because Bailey was a firm believer in democracy, in the idea that the "folk" (or demos) set the direction of government, he believed the most important key to improving rural life would be education—a specific kind of broad-based education that would prepare the country boy or girl to gain a deeper satisfaction from life while also becoming locally and civically engaged citizens who would represent the best interests of their communities.

"As with all education," Bailey wrote, nature study's "central purpose is to make the individual happy." In his characteristically playful but profound writer's voice, he continued:

> The happiness of the ignorant man is largely the thoughts born of physical pleasures; that of the educated man is the thoughts born of intellectual pleasures. One may find comradeship in a groggery, the other may find it in a dandelion; and inasmuch as there are more dandelions than groggeries (in most communities), the educated man has the greater chance of happiness.[2]

Such happiness and human flourishing, the kinds that empower adults to join children in finding delight in the wonders of a common dandelion, would emerge through direct contact with the earth and sympathetic understanding of the natural world. The nature study movement in the rural schools formed the base of this educational effort, and building from that would be the land-grant agricultural colleges, where farm kids could go to deepen their knowledge of farm life and return to their communities as leaders. This was also where experiment stations and extension agents would seek the cooperation of a given state's farmers to build the collective knowledge of that state for further dissemination and education. It is no wonder that Bailey became a leader in the early development of the agricultural colleges, eventually becoming the founding dean of the New York State College of Agriculture at Cornell University, a position he would hold for ten years, while his lectures continued to become some of the most legendary and popular around campus at that time. It is also no wonder that, in 1908, when President Theodore Roosevelt appointed a national Commission on Country Life to investigate the rural problems facing the nation, he turned to Dean Bailey to serve as the commission's chair.

As we were leaving the museum, John reached over to the museum's gift table and picked up two copies each of Richard Louv's *Last Child in the Woods* and Bailey's *The Holy Earth*. "Since no one else is here, and since I'm the director anyway," he said, "I want you two to have these." We thanked him, and he told us to keep in touch about the outdoor learning center. We told him we'd be back.

ULTIMATELY, IT WAS gardening, that most primal and personal practice of care that so directly links individual humans to the holy earth at their feet, that remained utterly central to all of Bailey's philosophical, educational, and scientific work. He was trained in botany, after all, and today he is perhaps best remembered as the "Father of Modern Horticulture," author and editor of such works as *The Standard Cyclopedia of Horticulture*, *Hortus* (the first horticultural dictionary), and *The Gardener's Handbook*, all of which can still be found on the bookshelves of farmers, nurserymen, and gardeners to this day. In 1885, at the age of twenty-seven and at the tail end of two years working in the Harvard herbarium of America's foremost botanist, Asa Gray, Bailey forever shook up the academic study of plants when he appeared before the Massachusetts State Board of Agriculture to deliver a legitimately stirring speech that he titled "The Garden Fence," imploring that the imaginary fence separating serious botany from the plants of the farm and garden be torn down. It was likely shortly before giving that landmark speech that Bailey excitedly informed Gray, his mentor at Harvard but also his childhood hero, whose *Botany* Bailey had carried around as a young man to help identify the plants he collected in the dunes and forests around his hometown, that he had been offered the opportunity to chair the nation's first Department of Horticulture and Landscape Gardening at his alma mater in Michigan. Gray did not share Bailey's excitement: "But, Mr. Bailey," Gray protested, "I thought you planned to be a botanist." "Yes, Dr. Gray," Bailey would remember responding, years later, "but a horticulturist needs to be a botanist." "Yes," responded Gray, "but he needs to be a horticulturist, too." When fellow Harvard botanist John Merle Coulter heard the news, he told Bailey flatly, "You will never be heard from again."[3]

To Gray and Coulter, and to the academic establishment generally, botanists were scientists, and horticulturists were merely gardeners. To Bailey, however, science and academic study ought to be coextensive with lived experience—and relevant to it. A single citizen with a seed packet and a garden plot, or, lacking that, even "one plant in a tin can," was simultaneously a scientist and an artist. Bailey knew that the digging and the watering and the weeding he did at the age of five in that small garden of pinks in front of his childhood home was science, and science of the highest order— open-minded, driven by curiosity, and capable of inspiring a lifetime of happiness—and he knew that it was art, full of expression and personality, too. From that first humble plot of pinks onward, every year of his ninety-six-year life Bailey found a way to keep a garden. Whether renting a plot of land in Shanghai while carrying out a series of studies in China, or tending year by year the ever-changing garden beds at his home in Ithaca, New York, and on the farm he managed in the summers, this was one point on which he would make no compromise. He needed his hands in the soil. And the result of that insistent connection to the earth is a lifework, including a seventy-six-book corpus, that maintains a remarkably stable and coherent, if ever-evolving, outlook on life and the world.

ABOUT A YEAR ago I visited John at his home in Otsego, Michigan. As he led me around his backyard, pointing out the trellises that the soft green tendrils of tomato plants were beginning to climb, the cucumbers and squash that were beginning to open their spicy, voluptuous yellow flowers, and, of course, the patch of pinks, I was reminded of the day eight years earlier when he had led my mother and me around the grounds and gardens of the Bailey Museum. Some of the museum's gardening experiments had clearly found rootage in his own yard since then. "And you know, man," he said, running his fingers through his sparse gray hair and scanning his own raised beds and mulched plots, "I never really grew anything before I started reading Bailey."

John stepped down from the museum's directorship in 2012 to spend more time at home in the summers with his wife and two kids. One side bonus has been the extra time to develop his garden. But just as Bailey's writings continue to shape John's life, they have continued to ripple through his into the lives of many others. The summer after that influential first tour of the Bailey Museum, I came back as an intern, beginning several years of working with John at the museum that eventually led to a couple years working as a transitional executive director after John stepped down. Bailey's quiet wit and wisdom about the beauty of growing things have continued to resonate in my life since then.

Meanwhile, the Liberty Hyde Bailey Outdoor Learning Center continues to grow at North Shore Elementary under my mother's inspired leadership. Without a dollar of aid from the financially strapped public school district, Mom's ad hoc volunteer corps of local landscapers, master gardeners, Girl Scouts and Eagle Scouts, and enthusiastic parents has laid out two trails through the woods, constructed a wetland overlook and a full outdoor classroom of moveable wooden seating, built and planted three raised beds, and, most recently, established a large butterfly garden planted with milkweed and other native flowers. The raised beds include a weed garden for hands-on learning and play (prominently featuring some beautiful dandelions), a bulb garden laid out as a grid to practice measurement where the kids can do more harmless digging around, and a perennial sun garden to observe the interactions of insects with a number of native flowering plants. Never are practicality, inspiration, or educational value sacrificed for each other—they all grow together with a symbiotic joy. Every day Mom takes her fourth-grade students on a run around one of the nature trails, several times each year a man from the Kalamazoo Audubon Society comes to take the kids birding, and several years ago a butterfly expert helped Mom get the class started raising and releasing monarch butterflies every spring and fall. My mother utilizes that acre or so of outdoor space to teach everything from poetry writing to the study of organisms, from physical fitness to the measuring of area and perimeter. Outdoor, hands-on learning is about much more than biology—as Bailey wrote, "Nature-study is not science. It is not fact. It is spirit. It is concerned with the child's outlook on the world."[4] And when that spirit of curiosity, interest, and love for the world spreads into a community, it catches—like wildfire.

I HOPE THE readers of this book catch that spirit themselves. This collection is, in many ways, the fruit of over a decade of John Stempien's journey, tracking paths that I have had the pleasure to share with him for the past eight years through Bailey's many writings, most of which are out of print. And it represents a welcome companion to Zachary Michael Jack's 2008 anthology of Bailey's written work, entitled *Liberty Hyde Bailey: Essential Agrarian and Environmental Writings* and also published by Cornell. The essays and poems in *The Liberty Hyde Bailey Gardener's Companion* provide an important dimension of the personal core at the center of the grander philosophical works in Jack's collection and complement perfectly the more personally reflective of Bailey's agrarian writings there. They also demonstrate some of the grounded, practical ways to apply the expansive vision that Bailey laid out in his full-length philosophical works, like *The Holy Earth*, which was recently brought back into print for a 2015 centennial edition with a new foreword, very worth reading, by Wendell Berry, available from Counterpoint Press. Liberty Hyde Bailey was a towering figure on the agrarian side of twentieth-century environmental thought, and his "biocentric" "earth-philosophy,"[5] which was so formative to Aldo Leopold's much later formulation of a "land ethic" and which continues to appear in the writings of such modern agrarian giants as Wes Jackson, Fred Kirschenmann, and Norman Wirzba, as well as Wendell Berry, should be more familiar to scholars of environmental philosophy, history, and literature than they are today.[6] But while scholarship catches up, John and I share the simple hope that we can all enjoy and find instruction in some of the most literary of Bailey's writings as they are presented in this new collection.

As John explains in his introduction, this anthology does not bring together Bailey's nuts-and-bolts, step-by-step horticultural guidelines or his more scientific botanical treatises of a century ago—while these writings remain instructive, an updated guide may be more useful—but focuses instead on what really made

Bailey the "Father of Modern Horticulture" in the eyes of so many readers over the decades: his nimble pen, his spirit, and his inimitable capacity to spread the gospel of gardening to every reader he reached. John and I have also explored in some more depth the ways in which the writings in this book unpack Bailey's larger philosophy, as well as some of the ways that certain selections have inspired and ignited individuals and groups, in the "Explanation of the Sections" following John's introduction. Bailey's impact on the history of popular engagement with environmental thought in the United States is often overlooked, but it seemingly can't be overstated—Bailey the journalist, like Bailey the administrator and the public intellectual, was a powerful, mobilizing force. It's that fire to grow things out of the good earth for the sake of growing them that we likely need now more than ever.

So take this book outside with you, read a chapter, and then go find a dandelion to admire. The point is to have fun and to find some life to care for, whether it be in the form of a dandelion, a pigweed, a pink, or a spear of asparagus. "It is often said that a weed is a plant out of place, but this is not so," Bailey wrote in 1927, at the age of sixty-nine. "Nothing is more in place than a weed!"

And as he wrote decades earlier in his delicious essay "The Dandelion," "Where once were weeds are now golden coins scattered in the sun, and bees reveling in color; and we are happy!"

John Linstrom
Brooklyn, New York, 2018

Acknowledgments

This book has been, in many ways, some twelve years in the making. The number of people who have lent us their support, inspiration, and guidance during this process are too many to list. Both editors offer their deep and abiding thanks to Kitty Liu and the team at Cornell University Press for making this dream a reality; to the incredibly helpful and friendly archivists and staff at the Cornell University Rare and Manuscript Collections; to Kendra Millis, who compiled the index, and to the Department of English at New York University for enabling us to hire her; to John Bailey Page, for his support of this project and for his permission to publish several essays and images, the inclusion of which he dedicates in loving memory to his grandmother, Annette Sailor Page; and to the team of volunteers and board members at the Liberty Hyde Bailey Museum in South Haven, Michigan, who got us started by encouraging and supporting our work on Bailey for nearly a decade. Last, we wish to express our appreciation for Liberty Hyde Bailey:

> If there be things I cannot tell
> The more I trust that all is well.

John Stempien

Just as a plant is dependent on specific conditions to germinate from a seed, this book too required its own unique support to blossom. I would like to express my gratitude to Dr. Kristin M. Sylvian, Elaine C. Stephens, and Jane L. Taylor, all of whom were integral in introducing me to the life, work, and spirit of Liberty Hyde Bailey. Personally, I would like to thank John Linstrom, who helped in the encouragement and envisioning of this collection's potential. I would also like to extend thanks to Paul English for his technical support during our sons' hockey practice. I would like to recognize my late mother, Jeanette Stempien, as well as Don Raiche, who both provided the invaluable model of "nature wisdom" through their interactions with the plant world. Finally, I thank my family, Emma, Josh, and Michelle, for their love and support.

John Linstrom

Thanks first to my parents, Robert and Rebecca, for their unconditional support even of projects that have seemed far-fetched and a bit distant from the task at hand. I have always benefited from your patience, as I have flourished under your loving guidance. Thanks too to Ben and Sam, without whom I would be both a lesser and a less interesting human, and probably less able to find myself in improbable situations. An unmeasurable well of gratitude belongs to the community of Baileyphiles whom I have been privileged to meet over the years, clustered around Cornell University and scattered elsewhere across the country, for their generosity in providing direction to a graduate student in no way affiliated with them on paper: Kevin Armitage, Ed Cobb, Bill Crepet, Bob Dirig, Elaine Engst, Peter Fraissinet, Wes Jackson, Fred Kirschenmann,

Kevin Lowe, Paul Morgan, Scott Peters, Daniel Rinn, Jane Taylor, and Gil Waldkoenig, among others. You have each been a model for me of unflinching academic integrity and unselfishness. Wendell Berry, the three hours I spent with you, when you showed me your marked-up copy of *The Holy Earth* and took me for a ride around the farm, were quietly but truly life-changing, and I will carry that visit with me for the rest of my life. For their understanding and leniency with an easily distracted doctoral candidate, as well as for their ongoing support and guidance in my academic life, I offer my deep gratitude to Una Chaudhuri, Sonya Posmentier, and Simón Trujillo. Thanks also to Mary Swander, my dear mentor and the creative force behind AgArts, and to the staff of Comstock Girl Scout Camp, for hosting me as their inaugural Bailiwick Writer-in-Residence, through the AgArts Farm-to-Artist Residency Program, at Bailey's summer home on the site of his old farm on Cayuga Lake. That time was inspirational and essential, and Mary's encouragement has fired up my Bailey work for close to a decade. For first introducing me to the joy and spice of Bailey's vast corpus, I gladly offer my deepest thanks to my first and primary mentor in the swirling universe of Baileyana, now my good friend and the man with whom I am honored to be coeditor of this volume, John Stempien. The journey has been collaborative; we have made good jazz; but the seed, for me, came from you, and you watered it. And finally, Monique, your enthusiastic support has sometimes been baffling and overwhelming to me, and it has always been incredibly gracious and nourishing. I love you, you fill my life with marvels, and I look forward to many years of tending the garden with you.

*The Liberty Hyde Bailey
Gardener's Companion*

Introduction

This book's seed was planted in 2006 in South Haven, Michigan, at the Liberty Hyde Bailey Museum, at one of its first weekly series of programs, which we called Brown Bag Botany. These were free summer lunch presentations, given on warm, halcyon days behind the white clapboard homestead of Bailey's parents, just about a mile from Lake Michigan. They took place underneath a grand, mature black walnut tree planted when the surrounding land was still the old Bailey farm. Never lacking for content, Bailey's instructional books in the museum library were an easy means to create easy programs. This was my introduction, as the museum's first director, to his writings, which, despite being removed nearly a century from our own time, were readable, engaging, and, frankly, a hit. His voice spoke. Local folks, sitting around picnic tables under the old tree, would nod along to passages, laugh at moments of Bailey's sparkling wit, and sometimes be moved to applause. Museum newsletters featuring his work would lead to museum blog posts, which in turn would lead to small in-house reprints of out-of-print material. So began a natural process to bring together for the first time an anthology of Liberty Hyde Bailey's most inspirational garden writings, which evolved into a labor of love to track down as many such pieces in Bailey's massive oeuvre as possible.

Chronologically, "The Garden Fence" is the oldest (and most academic) in this anthology, an 1885 lecture "read at the country meeting of the Massachusetts state board of agriculture at Framingham."[1] Arguably, without "The Garden Fence," Bailey's horticultural manifesto, none of the other writings in this collection—all of which were written with the everyday gardener in mind, or the aspiring one, whether gathered with friends under the old walnut tree or taking a break on the porch after a good afternoon spent digging in the side yard—could have been written. The timeless spirit that is evident throughout these selections first moved Bailey to stake his future on the then-controversial position for "a new horticulture" in which science and gardening would be compatible—and, perhaps more importantly, in which science and sentiment would be equally compatible—and a new literature based on this seemingly radical idea was needed to bridge the gap.[2] Simultaneously a serious scientist and a true poet, Bailey melded these seemingly disparate elements in his personality and his life. These two outlooks nourished each other, and we are fortunate that Bailey nurtured both; otherwise his writings, if purely scientific but not poetic, would be added to the historical dustbin of outdated gardening techniques. He also rooted his new horticultural writings in a rich legacy, interweaving his words and branching out his ideas with the reflections of past writers and poets: the pioneer horticulturist Patrick Barry, the eighteenth-century nature poets William Cowper and James Thomson, and the more notable—Charles Darwin, Ralph Waldo Emerson, Henry David Thoreau, and even Virgil. And of course there are the plants themselves, which he warmly refers to as "old friends." We are benefactors of the goodly heritage captured by Bailey's poetic insights into everyday beauty, which in turn point to that larger heritage of the earth that awaits us all in our own backyards.

Such a collection, surprisingly, has never been published. In this volume we have collected extracts from out-of-print books, uncollected essays that previously appeared only in farm journals and periodicals (many of which have never been digitized), and even previously unpublished essays like "Gardening and Its Future," which was read by Bailey over nationally broadcast radio in 1930 and which exists today only as a handwritten manuscript in the archives of Cornell University. As stated, these represent Bailey's literary, rather than practical, garden writings, and we have organized them thematically. Further elaboration on the organization of each section follows this introduction.

While we have attempted to group the selections in meaningful and enjoyable clusters, each passage stands independently of the others as well, and readers may either read them through in the order suggested here or explore them at their leisure as individual meditations. For this anthology we have kept the original spelling, punctuation, and capitalization, not only to retain the historic aspect of the text (e.g., "pease" instead of "peas") but also to preserve Bailey's literary style (e.g., coining new words and ideas through hyphenation).

If you are thinking of planting a garden, you will have your first garden after reading Bailey. If you are a novice, your vision will be expanded. If you are a well-seasoned gardener, you will find affirmations. As Bailey knew well, "Humanity began in a garden."

John A. Stempien
Otsego, Michigan, 2018

Explanation of the Sections, by the Editors

The first two section titles in this book come from the essay "To One Who Hath No Garden," in which Bailey insists that "there are two parts to gardening,—the growing of the plants in the soil, and the garden in the mind." As Bailey notes elsewhere, "The best gardener is the one who does the most gardening by the winter fire."[3] Section 1, then, titled "The Garden in the Mind," might represent the gardener's first considerations during the planning season, "by the winter fire," whether the desire is fruits and vegetables, flowers, or (more importantly) the "communion of the soul with the great mysteries of nature." As there are many kinds of gardeners, so there are many types of gardens, but the "garden sentiment" nurtured in these months is shared in common with all garden lovers, whether one works an acre or "one plant in a tin can." The reader can also envision the pieces in this section as representing what we might hear if we were to go knocking on the door at Bailey's home in Ithaca for a friendly hour's worth of sage advice. Such a scenario speaks even more aptly to the essay "Gardening and Its Future," written actually as a speech that Bailey read aloud over the radio in 1930, in the early days of the National Broadcasting Company (NBC), as the first in a series of nationalized radio pieces sponsored by the American Seed Trade Association. It was Bailey's first time "on air," bringing his voice into homes across the country, and this collection presents the historic speech in print for the first time. "The Garden in the Mind" ends, as all sections in this collection do, with a reflective poem from Bailey's poetry book *Wind and Weather*. The poem here, "Undertone," provides a glimpse into Bailey's famously vigorous work ethic, but it values more highly the "all-silent stream" of the individual's personality, artistically expressed through gardening, through poetry, and through life. The prose piece corresponding to this idea of an "undertone" or "background" to the common day will be found in the essay "Blossoms."

"The Growing of the Plants," section 2 of this book and the second "part" of gardening according to Bailey, could also be rightly named "Encounters in the Garden," since it focuses on the field of action. While these essays more directly engage issues of practical application ("How to Make a Garden—The First Lesson," "How to Make a Garden—Digging in the Dirt," "The Principles of Pruning," etc.), it is also here that we encounter Bailey as the seasoned botanist who desired for the gardener to "botanize intelligently in his garden"—not to dryly botanize as if for a stuffy museum collection, but to lean into taxonomy with the joy of someone coming into closer and more familiar contact with the plants as they grow. Bailey models this for us when he directs our gaze to the miracle of a "particle as dry and brown as a grain of sand" metamorphosing into stems, leaves, and "crimson flowers," and he models it in a different way as he takes us on a descriptive tour of his "specimen garden," featuring "307" unique plants from all over the world growing in a "vast and pleasant democracy." Do we dare to replicate his project? The more important point, for him, is that we fearlessly plant the garden of our dreams, no matter how eccentric, and irrespective of what anyone else might think. Plant not for your neighbors, but for yourself. For Bailey, a knowledge of gardening "means more than mere information of the plants themselves." It also includes a "sensitive mind," one that tunes itself to what is "nearest at hand," be it the vagaries of weather or the often-maligned weed. To "botanize intelligently" means all of this—it means an engagement with the emotions, the spirit, and the sensitivities of the individual, as well as the intellect.

From planning to planting: in section 3, "Flowers," the reader now arrives at the flowering stage of the plant, and hopefully, with that, the flowering of the garden sentiment. One of Bailey's favorite garden writers seems to have been the English farmer-muckraker William Cobbett (1763–1835), whom he describes as "original" and "pungent" in the opening essay of this collection.

We might use the same words to describe Bailey's reflections on flowers in this section—certainly the perspective he invites us to take would qualify as "original," and it challenges the competitive garden-club mentality that sometimes takes over the garden sentiment and turns it superficial:

> I dislike to hear people say that they love flowers. They should love plants; then they have a deeper hold.

> I have sometimes wondered whether the average flower-buyer knows that flowers grow on plants.

> The greatest fault with our flower growing is the stinginess of it.

> Nature has no time to make flower-bed designs. She is busy growing flowers.

The garden sentiment that flowers in this section, then, defies convention and invites the gardener into a new kind of relationship with the things that grow from the good earth. Flowers represent, in Bailey's worldview, much more than the beauty that they physically display—they bring us in, like bees to the nectar, but they can also lead us to something more profound: a sense of kinship with the whole plant that "lies deeper even than the colors, the fair fragrances, and the graces of shape. It is the joy of things growing because they must, of the essence of winds woven into a thousand forms, of a prophetic earth, and of wonderful delicateness in part and substance. The appeal is the deeper because we cannot analyze it, nor measure it by money, nor contain it in anything that we make with our hands. It is too fragile for analysis." Bailey beautifully lays out this depth of influence in the section's opening essay, "Blossoms," and the following selections all serve to expand upon his philosophy of flowers.

Section 4, "Fruits & Vegetables," contains some of Bailey's more autobiographical essays, connecting to his formative years on his family's fruit farm in South Haven, Michigan. The epigraph that opens the section, three essays, and the poem that closes the section originate from a seven-book series titled The Background Books: The Philosophy of the Holy Earth. The Background Books would become Bailey's most significant foray into environmental philosophy, and in them, Bailey's childhood becomes a touchstone to reflect on the beauty and mystery of the processes from flower to fruit, whether the example be that of apples, peaches, or the orchard itself. The growing of food for necessary sustenance was a reality of Bailey's young life on the frontier farm, and that agrarian experience fundamentally shaped his philosophy. It embodies a dependence on the produce of the garden that is even more fundamental than the spiritual and aesthetic dependence on flowers grown for ornament. Even the "one who hath no garden" depends upon the "fruits and vegetables" grown by others to live and to flourish, and the kitchen gardener knows this intuitively. These essays, then, reflect a philosophical maturity akin to the maturity of the fruit-bearing plant.

Section 5, "Spring to Winter," tracks the rhythms of nature as they are expressed to sensitive gardeners, to those who have experienced firsthand the care for growing things, have internalized and processed the philosophy of that care, and then think back, perhaps again "by the winter fire," on the chapters they have watched unfold from conception to planting to stewardship to harvest. We must imagine that in Bailey's own fireside ruminations, these domestic memories flowed seamlessly into more exotic ones, because, while Bailey was a great lover of the home and garden, he was also a citizen of the world, a passionate traveler, and a plant collector. Yet despite over a hundred major plant-collecting expeditions over the course of his long life (the last being a solo trip to the Caribbean at the age of ninety-two), many of them to the tropics in his effort to detangle taxonomic confusion over the Palm family, Bailey noted an "inexpressible deficiency" in his experience of tropical surroundings. That deficiency was expressed, in "Where There Is No Apple Tree," through the absence of his most beloved species of fruit tree and the rhythm of the seasons he attached to it; in the tropics, as a foreigner and in spite of his training as a botanist, Bailey lacked that connection to the local environment. In an essay in this section, "An Outlook on Winter," he writes, "Many times in warm countries I have been told that the climate has transcendent merit because there is no winter. But to me this lack is its disadvantage." For Bailey, the cycle of the seasons possessed transcendent merit, because, in this cycle, "there is no ending and no beginning, only stages in a persisting and everlasting process." And as he so beautifully expresses in the poem "Apple-Year," each of those seasons, through the sunshine, dew, and frost, is contained in every individual apple—the garden brings the seasons home. Here we see Bailey the naturalist, exalting in each season and the gifts that each affords, from the overlooked spring dandelion to the winter grass pinks with their blue-green tufts peeping through the snow. These essays serve to bridge the act of gardening to what Bailey coined the "everlasting backgrounds," the fundamental spaces that help us in the process of "connecting everything" and reveal the "permanent and invariable attributes in our occupancy of the Earth"—after all, "the backgrounds are the realities," and in our sometime forgetfulness we must work every day to transform them from apparent backgrounds into the *foregrounds* of our consciousnesses.[4] For Bailey, growing plants can awaken us to this task.

The first piece in our two-part epilogue, "Marvels at Our Feet," first published in *The Land Quarterly* in the 1940s, serves as a kind of anthology in itself, excerpting and weaving together portions of some of Bailey's most memorable essays from his long career as an author. You may recognize some portions from essays found elsewhere in this collection. The essay first appeared between the covers of a book in 1950, in a collection of the best writings from *The Land*, just four years before Bailey's death at the age of ninety-six. We might think of it as Bailey's last words, an overarching statement that sums up his life's philosophy with grace and poetry. It stands out as a remarkable composition for a man in his late eighties. It is a rich synthesis.

But, for a final word in this collection, we leave you with the stirring conclusion of another book, *Universal Service*, published in 1918 for the Background Books series. "Society of the Holy Earth" calls for a new kind of organization, a sort of grassroots movement, one of "few officers and many leaders . . . controlled by a motive rather than by a constitution. . . . Its principle of union will be the love of the Earth, treasured in the hearts of men and women." If the practice of gardening provides the foundation of Bailey's "earth-philosophy," such spontaneous coming together of stewards of the planet represents that philosophy's ultimate social expression. Bailey envisioned something much larger than a formal organization—it would have to become a movement, one that could reach people of all backgrounds and walks of life—and this is evidenced by the way in which he revisited the passage many years later. When Bailey was sent a copy of the first issue of *The Land Quarterly* in 1940, he wrote a letter to the editor in praise of the new journal and its newly formed sponsoring organization, The Friends of the Land. In that published letter, he invites the journal's readership to revisit his suggestion, of 1918, for a

"Society of the Whole Earth." Bailey, by then known as "the Sage of Cornell," had been visited by the founders of the organization in the wake of the Dust Bowl as they sought advice on how best to organize around the cause of soil conservation. The young group of idealists had chafed against Bailey's proposed title, joking that the language of "holiness" might cause some people to mistake the group for some kind of cult. Despite Bailey's conviction that the earth was indeed holy (for the simple reason that "man did not make it," as he writes in *The Holy Earth*), he also knew that his vision could transcend the limitations of creed and dogma, so in the letter that he sent them for publication in their fledgling journal (a journal that would go on to publish work by the likes of E. B. White, William Carlos Williams, Louis Bromfield, Aldo Leopold, and even a young Rachel Carson), Bailey gave the name of his envisioned society a gentle enunciative nudge, from *holy* to simply *whole*.[5] The earth forms a whole in Bailey's thought, one that radiates out from the individual organism through the interconnectedness of all life. The microcosm of this vast fellowship would be found most poetically and personally, Bailey believed, in the garden. In the end, he knew that the keepership of the planet must be a collective enterprise, motivated by the love of things that grow and nurtured in the soil of our own backyards.

I

The Garden in the Mind

Like the love of music, books and pictures, the love
of gardens comes with culture and leisure and with
the ripening of the home life. The love of gardens,
as of every other beautiful and refining thing, must
increase to the end of time. More and more must
the sympathies enlarge. There must be more points
of contact with the world. Life ever becomes richer.
Gardening is more than the growing of plants: it is the
expression of desire.

General Advice

Every family can have a garden. If there is not a foot of land, there are porches or windows. Wherever there is sunlight, plants may be made to grow; and one plant in a tin can may be a more helpful and inspiring garden to some mind than a whole acre of lawn and flowers may be to another. The satisfaction of a garden does not depend upon the area, nor, happily, upon the cost or rarity of the plants. It depends upon the temper of the person. One must first seek to love plants and nature, and then to cultivate that happy peace of mind which is satisfied with little. He will be happier if he has no rigid and arbitrary ideals, for gardens are coquettish, particularly with the novice. If plants grow and thrive, he should be happy; and if the plants which thrive chance not to be the ones which he planted, they are plants nevertheless, and nature is satisfied with them. We are apt to covet the things which we cannot have; but we are happier when we love the things which grow because they must. A patch of lusty pigweeds, growing and crowding in luxuriant abandon, may be a better and more worthy object of affection than a bed of coleuses in which every spark of life and spirit and individuality has been sheared out and suppressed. The man who worries morning and night about the dandelions in the lawn will find great relief in loving the dandelions. Each blossom is worth more than a gold coin, as it shimmers in the exuberant sunlight of the growing spring, and attracts the bees to its bosom. Little children love the dandelions: why may not we? Love the things nearest at hand; and love intensely. If I were to write a motto over the gate of a garden, I should choose the remark which Socrates made as he saw the luxuries in the market, "How much there is in the world that I do not want!"

I verily believe that this paragraph which I have just written is worth more than all the advice with which I intend to cram the succeeding pages, notwithstanding the fact that I have most assiduously extracted this advice from various worthy but, happily, long-forgotten authors. Happiness is a quality of a person, not of a plant or a garden; and the anticipation of joy in the writing of a book may be the reason why so many books on garden-making have been written. Of course, all these books have been good and useful. It would be ungrateful, at the least, for the present writer to say otherwise; but books grow old, and the advice becomes too familiar. The sentences need to be transposed and the order of the chapters varied, now and then, or interest lags. Or, to speak plainly, a new book of advice upon handicraft is needed in every decade. There has been a long and worthy procession of these handbooks,—Gardiner & Hepburn, M'Mahon, Cobbett—original, pungent, ubiquitous Cobbett!—Fessenden, Bridgeman, Sayers, Buist, and a dozen more, each one a little richer because the others had been written. But even the fact that these books pass into oblivion does not deter another hand from making still another venture!

I expect, then, that every person who reads this book will make a garden, or will try to make one; but if only tares grow where roses are desired, I must remind the reader that at the outset I advised pigweeds. The book, therefore, will suit everybody,—the experienced gardener, because it will be an echo of what he already knows; and the novice, because it will apply as well to a garden of burdocks as of onions.

A garden is the personal part of an estate, that area which is most intimately associated with the private life of the home. Originally, the garden was the area inside the enclosure or lines of fortification, in distinction to the unprotected area or fields which lay beyond; and this latter area was the particular domain of agriculture. This book understands the garden to be that part of the premises which is devoted to ornament, and to the growing of vegetables and fruits either for the home consumption or for market. The garden is, therefore, an ill-defined demesne; but the reader must not make the mistake of defining it by dimensions, for one may have a garden in a flower-pot or on a thousand acres. In other words, this book believes that every bit of land which is not used for buildings, walks, drives and fences, should be planted. What we shall plant,—whether sward, lilacs, thistles, cabbages, pears, chrysanthemums or tomatoes,—we shall talk about as we proceed.

The only way to keep land perfectly unproductive is to keep it moving. The moment the owner lets it alone, the planting has begun. In my own garden, this first planting is of pigweeds. These are usually followed, the next year, by ragweeds, then by docks and thistles, with here and there a start of clover and grass; and it all ends in June-grass and dandelions. Nature does not allow the land to remain bare and idle. Even the bank where plaster and lath were dumped two years ago is now luxuriant with burdocks and sweet clover; and yet people who pass that dump every day say that they can grow nothing in their own yard because the soil is so poor! Yet, I venture that those same persons furnish most of the pigweed seed which I use on my garden.

The lesson is that there is no soil,—where a house would be built,—so poor that something cannot be grown. If burdocks will grow, something else will grow; or if nothing else will grow, then I prefer burdocks to sand and rubbish. The burdock is one of the most striking and decorative of plants, and a good piece of it against a building or on a rough bank is just as useful as some plant which costs money and is difficult to grow. I had a good clump of it under my study window, and it was a great comfort, but the man would persist in cutting it down when he mowed the lawn. When I remonstrated, he declared that it was nothing but burdock; but I insisted that, so far from being burdock, it was really Lappa major, since which time the plant has enjoyed his utmost respect. And I find that most of my friends reserve their appreciation of a plant until they have learned its name and connections.

"The ornamental burdock."

The dump-heap which I mentioned has a surface area of nearly one-hundred and fifty square feet, and I find that it has grown over two hundred good plants of one kind or another this year. This is more than my gardener accomplished on an equal area, with manure and water and a man to help. The difference was that the plants on the dump wanted to grow, and the imported plants in the garden did not want to grow. It was the difference between a willing horse and a balky one. If a person wants to show his skill, he may choose the balky plant: but if he wants fun and comfort in gardening, he had better choose the willing one.

I have never been able to find out when the burdocks and mustard were planted on the dump; and I am sure that they were never hoed or watered. Nature practices a wonderfully rigid economy. For nearly half the summer she even refused rain to the plants, but still they thrived; yet I staid home from a vacation one summer that I might keep my plants from dying. I have since learned that if the plants in my borders cannot take care of themselves for a few weeks, they are little comfort to me.

To One Who Hath No Garden

There are two parts to gardening,—the growing of the plants in the soil, and the garden in the mind. The desire to have a garden comes first; then comes the season of planning, the pleasant discussion of the kinds, the tools, the construction of hotbed and frame, and the layout worked over and over again until the area, the desired products, and the purse are all accommodated and made to fit; finally comes the putting of the plan into execution.

I know persons who are musicians and yet have no musical instruments. Some of them can perform on instruments and some of them cannot. If they are performers, they miss the instruments more. Do not most of us, with high taste for music, secure our satisfaction in it from those more fortunate or more skillful than we?

I know poets who do not write poetry, artists who do not paint, architects who do not build. I know gardeners who do not garden.

It is not for me to depreciate the joy and value of a garden that one makes in the good earth with one's own hand; yet the garden is an appreciation. It is an appreciation of activity, of color, of form, of ground smells, of wind and rain and sun, of the day and the night, of the things that grow. Good critics of gardens, good lovers of gardens, may yet not be good gardeners; and good growers may not be deep appreciators of gardens.

To the one who has no garden (my sympathy is his!) there still remains some of the essential joys of the garden,—the wonders of the catalogues, the invitation of the soil, the discriminating knowledge of the plants. A garden is only a piece of the world,—a piece that one picks out and arranges for one's own exercise and pride. Beyond it are others' gardens, also the open greensward of fields, the wonderful trees, and flowers at one's feet, the voices of birds, and the abounding atmosphere. One may sit at another's garden gate, and feel its beauty; one may wander afield in any afternoon or holiday; one may be open to the suggestion of garden and beauty as one travels back and forth, missing nothing.

We wish that every person might have a garden. We wish also that every person might have an instrument of music or good books of verse. Yet the year is not lost without them. And if one has not a garden then must one make the most of the compensations, never foregoing the satisfactions in the gardens that others make, in the gardens kept by the public for such as they. This is only to say that we would have the garden sentiment possessed of all the people, missing not one; some of the people will grow their own gardens also.

The Common Natural History

The first consideration of special study should be the inhabitants of your yard and garden: they are yours; or if they are not yours, you are not living a right life. Do you wish to study botany? There are weeds in your dooryard or trees on your lawn. You say that they are not interesting: that is not their fault.

We have made the mistake all along of studying only special cases. We seem to have made up our minds that certain features are interesting and that all other features are not. It is no mere accident that many persons like plants and animals but dislike botany and zoölogy. It is more important to study plants than special subjects as exemplified in plants. Why does the weed grow just there? Answer this, and you have put yourself in pertinent relation with the world out-of-doors.

If one is a farmer, he has the basis for his natural history in his own possessions,—animals domestic and wild, plants domestic and wild, free soil, pastures and lowlands and woodlands, crops growing and ripening, the daily expression of the moving pageant of nature. Zoölogical garden and botanical garden are here at his hand and lying under his title-deed, to have and to hold as he will. No other man has such opportunity.

I would also call the attention of the townsman to his opportunity. If the range of nature is not his, he still has the wind and
rain, the street trees, the grass of lawns, the weed in its crevice, the
town-loving birds, the insects, and I hope that he has his garden.
Even the city has its touch of natural history—for all things in the
end are natural, and we recognize them if we have had the training
of a wholesome outlook to the commonplace. Timrod's sonnet on
the factory smoke is a nature-note:

> "I scarcely grieve, O Nature! at the lot
> That pent my life within a city's bounds,
> And shut me from thy sweetest sights and sounds.
> Perhaps I had not learned, if some lone cot
> Had nursed a dreamy childhood, what the mart
> Taught me amid its turmoil; so my youth
> Had missed full many a stern but wholesome truth.
> Here, too, O Nature! in this haunt of Art,
> Thy power is on me, and I own thy thrall.
> There is no unimpressive spot on earth!
> The beauty of the stars is over all,
> And Day and Darkness visit every hearth.
> Clouds do not scorn us: yonder factory's smoke
> Looked like a golden mist when morning broke."

The Importance of Seeing Correctly

Three professional fruit-growers expressed an opinion concerning the manner in which sweet cherries are borne on the tree. The first contended that the fruit grows from side spurs on twigs which grew last fall. The second was equally positive that it is borne on short spurs which grow from the point of junction between last year's wood and the wood of the year previous. The third supposed that cherries grow in pairs from most of the buds on both last year's and two years old wood. Each of these men had grown cherries for at least a dozen years, and yet neither of them knew this one of the simplest facts connected with their daily labor, and which might be made apparent by a few minutes' close observation. It is surprising that many of the commonest and most interesting of everyday phenomena, though they lie right before the eyes of every man, are never seen by the great majority of people. Most persons are walking through a wonderland with their eyes shut. The interesting things detailed in these pages are but a very few random leaves rudely torn from the book of nature. The leaves that remain are fully as inviting, and they are doubly profitable when Nature herself tells the story.

One needs practice, along with scientific training, to interpret aright all the things that he may see. A farmer of my acquaintance noticed that grasshoppers appear shortly after the stems of golden-rods become affected with peculiar frothy swellings, and he at once asserted that the grasshoppers bred in the golden-rods! If he had carefully cut open these swellings he could have found proof enough against his assertion. Another friend noticed that the long-stalked and therefore conspicuous flowers of his pumpkins had all died: he immediately proclaimed to his neighbors that his pumpkins were blasted, and that the entire crop in that vicinity would be small! Had he known that these flowers were staminate, and that when they had shed their pollen their mission was ended, he should have had greater wonder if they had not died. Still another friend discovered a minute insect boring into a pear-tree, and as that tree happened to be blighted he announced that a certain insect was the cause of pear blight; nevertheless, a score of other trees which had the blight would probably show no sign of the ominous insect. It is never safe to draw conclusions hastily, and especially not from one or two detached observations. I will relate a very sober incident, of which an account was published a short time since in an agricultural paper, and I request that my readers bear it in mind as an antidote against hasty conclusions. An observing fruit-grower possessed a plat of smooth-fruited gooseberries. A favorite family cat, having unceremoniously died, was buried underneath one of the gooseberry bushes, and behold! the next year that bush bore hairy berries, and has so continued to do unto the present day!

A Reverie of Gardens

If I could put my woods in song
And tell what's there enjoyed,
All men would to my gardens throng,
And leave the city void.

Into this quatrain Emerson has put the expression of a universal passion—the passion to know the fields and the growing things. This desire may express itself, as with Emerson, in a longing for the place where "the savage maples grow" and "no tulips blow," or in a yearning to break the earth and make a garden.

The nature-desire may be perpetual and constant, but the garden-desire returns with every new springtime. Recently an agitation for planting was begun in the city schools of Rochester. A local seedsman put up small packets of flower seeds in ten varieties at a cent a packet; in two weeks 11,000 packets were purchased by the children. With the first warm days of spring, the city resident goes forth with spade and hoe and fills the back yard with the tomato-plants that the enterprising gardener kindly placed on sale at the grocery-store. Frost kills the plants, but the amateur buys again, for the gardener has learned to keep up the supply of tender plants on the grocery stands. The vender of impossible rose-bushes finds a ready buyer. The tree-pruner, with occult knowledge and a secret remedy for all the ills of trees, discovers an easy client. It is the season of expectancy. Every bud is a promise. The soft, sweet-smelling earth is fat with possibilities. With a lavish hand the planter plants. The days are sweet and cool. The air is new and clean. But presently the days become pinched. The air is humming hot. The bugs grow to fatness. The weeds come. Dust settles over the herbage like a coat of ashes. One by one the plants smother and die. The enthusiasm of springtime is withering, and in the parching suns of August it is but a memory. Then come the sad ripe days of autumn. The mellow sunshine and falling leaves force one out of doors. Every nook and corner of the place is visited, for the winter is coming when the mystery of sleep will be on the garden. We snuggle the tender crowns over with leaves. We fill the beds with hardy bulbs. We see the last leaf fall. Next spring the old enthusiasm will burn again.

Does my reader recognize the picture? Does he not remember how often he has questioned whether, after all, it is worth the while to make a garden, because it is cheaper to buy his plants than to raise them? If so, then listen! The value of a garden, as of every other good thing, is in the pleasant impulse as much as in the final product. Just to have handled the clean new earth, and to have sown the seed, and to have thought about the garden at morning and at night—this is worth the effort. You have had a new experience. You have come nearer to nature. Next fall the Rochester children will make an exhibition of their plants. Some of them, favored by good soil or aided by brother or sister, will make a beautiful display. Others will look on silently and perhaps sadly, for they will have failed; but my heart will go out to these, for perhaps they will have gained as much as their comrades, if only they could know it.

It is an interesting fact that, with all the segregation of people into cities, the sales of seeds and plants are increasing immensely. It would seem that there is no limit to the number of things in which men and women can find interest. With all the increasing complexities of civilization and the wearing details of business, we are interested in a greater number of incidental and secondary things than our fathers were. The very stress of our lives is driving us to nature for respite. An interest in plants is part of our culture and our leisure, akin to the interest in books and pictures and music.

Sooner or later, every person feels this desire to plant something. It is the return to Eden, the return to ourselves after the long estrangement of our artificial lives. One of us dreams of a little patch of orchard bounded by cool, grassy banks. Another wants a snug and tidy garden-plat bounded by a wall and a lattice, and at one side a tinker's room of tools, rakes and hoes and watering-cans, and assorted sizes of pots, and boxes containing string and labels and screws and bits of wire. In this room he would work when the rain falls heavily on the roof and pours across the doorway from the wide-hanging eaves. Others want long, trim rows of strawberries, beets, and onions, with beds of lettuce, hills of squashes, and clumps of hyssop and sage in the corners, all ranged and labeled as the books on a shelf. Others want tumbling piles of vines shot through with wild asters and the spires of hollyhocks. Still others would roam afield and find their satisfaction in the things that by chance have found a place in which to grow. But, whatever the form of the wish, the substance is the same—it is the natural man longing to express itself. It is the desire to be alone with something that understands you. I have heard the gardener talk to his plants, and not one of them disputed with him.

To one person, this desire for the out-of-doors is an expression of the art-sense. To another, it is respite and release. To another, it is the joy of seeing and touching real things. To another, it is health and physical exercise. To others, it is natural history. To others, it is gardening or farming. To some, it is communion of the soul with the great mysteries of nature.

Have you made a garden all by yourself? Then try it, if you have not. Do not delegate the work. Yourself thrust the spade deep into the tender earth. Bear your weight on the handle and feel the earth loosen and break. Turn over the load. You smell the soft, moist odor, an odor that takes you back to your younger and freer days or sends you dreaming over the fields. You have uncovered the depths where the earthworm burrows and the pupa has lain since midsummer. Run your fingers through the soil. It is mealy and fine and clean. It may have been turned a hundred times, yet it is new and virgin. You feel as if you could plant your feet in the soil and grow like a plant. Spade up the whole bed. Note how the loose earth settles into place as you draw your rake back and forth. The moisture steams from its bosom, and the drying surface affords a mulch to hold the water that lies in its depths. You are wondering what is contained in this earth. Men have spent their lives to answer that inquiry and have died without making the answer complete. Compounds of potassium and phosphorus and silicon and nitrogen and many more; millions of micro-organisms that you cannot see, and whose lives no man knows; chemical activities too complex to be analyzed; moisture and heat and magnetism; physical forces so intricate and subtle that they cannot be measured—all these are in the laboratory that you are preparing. Yet you are powerless to bring them forth into visible life. One day you will drop a speck of matter into the soil, a speck so small and round that you must depend on the label to tell whether it is cabbage or turnip or cauliflower or mustard; and, behold! a new being comes forth, endowed with life, with roots and stem and leaves, and flowers and fruits and seeds, all unfolding in their appointed sequence and season! Where is the seat of this mystery, of the alchemy that makes one seed unfold into a turnip, and its counterpart into a cabbage? I often wonder how a cabbage-seed knows that it is a cabbage-seed.

You will see the fresh green heads of the perennial herbs pushing up the mold on April Fool's day. You will wonder whether they are bleeding-hearts or daylilies, or merely wild stray things without names. You will see the tiny new things grow. You will wonder at the shapes of the leaves. You will see the first flower-bud nestled deep in the heart of the foliage. You are discovering a new world, and the quest is all the pleasanter because you have made the same discovery every spring. The longer you know the plants the more they mean to you. You never tire of old friends. The joys of association are added to the interest in the plant itself. By and by, I hope, you will learn that great fact that so few people ever learn, the fact that a plant is a plant whatever its name or kind. I sometimes wish that we could transpose the names of plants; then would a violet be a dock and a dock would be a violet.

When a plant interests you because it is a plant, you have graduated in nature-wisdom. Then you will care less what are the form and fashion of your garden than that it contain plants. You will be interested in the plant itself quite as much as in its flowers or fruits. You will be interested merely to see what comes. I sometimes think that the amateur's gardening is a mild kind of gambling. Every seed is a lottery ticket. You are curious to know what will come out of it. I like to sow the mixed seeds that are left in the gardener's box, the accumulation of all the kinds that have fallen from the packets. You press the seed into the warming earth, then wait and watch. When the first tiny sprig appears, you are repaid for your effort, for this is the most interesting stage in the plant's life. I sometimes wish that the seed catalogues would tell how a plant looks when it first comes up, rather than the color of its flowers.

You will know when you like plants because they are plants—your interest in them will not depend on the flowers, nor even on the foliage. You will like plant forms. The leafless winter tree will appeal to you. The heads of teasel and the fretwork of wild carrot standing above the snow will bring you joy and peace. You will let the dry stalks stand in your garden that you may see them all winter; then will your garden be perpetual.

Of course your garden will be trim, orderly, and neat, at least when you begin. Such an arrangement indicates methodical habit and a purpose to succeed. Long straight rows at regular distances, plants graded as to height and kind and season, add greatly to the attractiveness and purposefulness of the garden, when the garden is looked upon as a piece of effort. It will have a look of good workmanship. It will afford endless opportunity for pleasant puttering in keeping down the weeds, training the plants, removing stray shoots, and in breaking the crust of the soil. It will be easy to tend and till. There is satisfaction in a trim border of foxgloves or hollyhocks.

The best garden will be at one side of the place, or at the rear. A flower-bed in the middle of the lawn is impertinent. The flowers are at a disadvantage as compared with the roots of the sod, for these roots have first levy on food and moisture; and a bed in the lawn spoils the lawn. You will delight to fill in the corners of the house with motherwort and other pleasant weeds. You will round off the corners with billows of trumpet-creeper. You will overhang the rear walk with vistas of foliage. You will scatter bright flowers in the border by the kitchen window. A garden or a plant is valuable for the place it occupies as well as for itself. There is satisfaction in the yard in which all parts blend and harmonize; it has character as a whole and as a picture. It has meaning. A yard that has individual plants scattered over it hails you as you pass; and each plant shouts, "See! I cost five dollars!"

Yet, with all this, who is not drawn to the neglected garden? What are the old-fashioned gardens that we love, if they do not have an old-time air of abandon and unrestraint? There is one that I visit often. It is one of the estates of a past generation, built when land was cheap. It has seen its day. Now there are rotting piles of wood grown over with moss and the weedy tangles of bittersweet. Old trees have fallen, and unfamiliar plants are growing about their prostrate trunks. In these early days of May the new grass is soft and thick. The old sod springs under your feet. A full coinage of dandelions is scattered on the grass. You scrooch under the broken trees and pick your way through thickets of plum-sprouts and lilac. Old-fashioned daffodils and jonquils and grape-hyacinths rise from the grass and mark the sites of former beds which in their prime had trim borders of flagstone or of box. In the open places are soft cushions of celandine, catnip, motherwort, and bluebells. Shy-faced sweet violets hide themselves in the deep turf. The wide spears of tulips are springing along the old walks, and amongst them the wild cleavers are clambering for sunlight. A wide-spreading clump of striped canary-grass marks the spot of some old flower-bed. From front to rear at one side of the place stretches the ruin of a promenade. A generation since a summer-house stood at its further end; the broken weather-gray lattice marks its place. The box-edging, in a double broken windrow, alone recalls the formal beauty of the design. Into the edging the plum-sprouts have intruded, and daffodils and crown imperials have worked their way into the nooks and spaces. Here one can botanize. Here the wild birds nest; and in the season one may hear the warblers as they stop in their long migrations, for the birds seem to have an instinct that leads them to neglected gardens.

But these old gardens interest us because of the memories that they recall, not alone because they are gardens. They are suggestive of human lives. Yet I fancy that more than one human being has been led to a love of plants from having first known them in some grandmother's garden. We would not make a neglected garden, for intentional neglect is not neglect of the neglectful kind; but neglect that comes naturally and easily will be no indication of failure if only we find satisfaction in the decline; and hereby am I the more willing to urge everyone to make a garden! A garden in which one finds joy cannot be successless.

There is no weather that does not suit some plant. In the hottest and driest time the portulacas are burning red on the sand. In cool and cloudy weather the soft morning-glories remain open until noon. When the soil is soaked with rain, the irises are in their glory. The plants have no dread of storm. Note the hang of the leaves and the droop of the flowers in the beating rain. Note that the chickens do not run, but throw their bodies back to shed the rain and then stand in philosophic comfort. Thoreau was glad when it rained because his beans were happy. A garden is the best of remedies for that commonest of melancholies, the habit of grumbling at the weather.

Your chief joy in your garden will not be in the vegetables that you eat nor in the flowers that you pick, but in the satisfaction of causing things to grow. You will enjoy the companionship of things that are real and clean. You will come to know the common and the little things. Some time, without knowing it, you will let a pigweed grow; and then you will be sorry to pull it up.

The Feeling for Plants

One does not make a good library till one has a feeling for books, nor a good collection of pictures without a feeling for pictorial art. Neither does one make a good garden of any kind without a feeling for plants.

Education in Terms of Plants

This does not mean that the feeling must be born with the person. It would be a hopeless world if we could not acquire new sentiments and enthusiasms. One can cultivate a feeling for plants by carefully observing them, growing them, reading about them, and particularly by choosing the company of persons who know and love them. As soon as one begins to distinguish the different kinds closely, one acquires the feeling of acquaintanceship; every kind then has its own qualities, and every kind is admirable in itself. Plants have personality.

The Forms of Plants

The interest in plants is primarily, I suppose, in their forms. They are endlessly diverse. Vine, herb, tree, shrub, aquatic, they inhabit the earth and clothe it, and give significance to scenery. The green-ery of vegetation is the mantle and the garnish of the planet. Leaf-forms, flower-forms, fragrances, shapes and colors and odors in fruits, twig-habit and bark and buds are all perfect of their kind. To admire a plant is to be keen in observation, appreciative of nature, responsive in sympathy and suggestion.

The plant-grower has a special intimacy with his plants. They respond to his care; they come up slowly from the seed or the cut-ting; they take on new forms and adapt themselves to the condi-tions he provides. Often will one see a gardener run his fingers over the stem or branches and pass his hand over the foliage as if caressing the plant.

The Beauty of Growth

The lover of plants enjoys them in their surroundings, in the places where they grow. When they seem to fit the place, or become a part of the general composition, they have the added beauty of association, one plant complementing another. The growth-form of one differs from the form of another; the color and fashion of bark are different; the foliage effects are distinct; yet they may not be inharmonious.

The plant-lover responds to the plants as they grow in the wild. The bush by the roadside interests him; he looks for it as he comes and goes. The fence-row has its charm, even though he must cut it out to make room for crops. The herbs and the trees, the plant-forms in the marsh, all awaken a pleasurable response. He wants to transfer them to his grounds.

It is well to have a nursery plot at one side, out of sight and out of the way, to which all kinds of things from the wild may be transferred. As they grow, some of them may be wanted for the grounds, and in any case, there is the pleasure of anticipation, of experiment.

From Month to Month

Much of the interest in plants is conditioned on the seasonal changes. In this are they unlike animals, and hereby do they have a special charm. The swelling of the buds in spring marks an epoch: the birds come back; the creeks are overflowing; a new odor rises from the earth; the sky is soft; the men and teams take to the fields. Then the buds burst, the leaves unfold and grow, the branches lengthen, the foliage is complete, the flowers come and fade, fruit appears; then comes the yellowing of the leaf, the dropping one by one as the autumn moves on, and finally the bare twigs go well prepared and secure into the great test of winter. Next year, will the miracle be repeated? We know it will!

The Growing of Plants

After a time one expresses one's knowledge and skill in the raising of plants. The kinds come to be familiar. The books and catalogues have a new meaning. Acquisitions are prized. Experiment is fascinating. One is proud of one's workmanship. Then does the growing of plants become a real enthusiasm.

No modern home that has a yard is meeting its best opportunities unless it exhibits a discriminating feeling for plants. One owes it to oneself to cultivate an appreciation of plants, of gardens, and of landscapes. One owes it to one's family and to the children.

Planting a Plant

Most persons are interested in plants, even though they do not know it. They enjoy the green verdure, the brilliant flower, the graceful form. They are interested in plants in general. I wish that every person were interested in some plant in particular. There is a pleasure in the companionship, merely because the plant is a living and growing thing. It expresses power, vitality. It is a complete, self-sufficient organism. It makes its way in the world. It is alive.

The companionship with a plant, as with a bird or insect, means more than the feeling of the plant itself. It means that the person has interest in something real and genuine. It takes him out-of-doors. It invites him to the field. It is suggestive. It inculcates a habit of meditation and reflection. It enables one to discover himself.

I wish that every child in New York State had a plant of his own, and were attached to it. Why cannot the teacher suggest this idea to the pupils? It may be enough to have only one plant the first year, particularly if the pupil is young. It matters little what the plant is. The important thing is that it shall be alive. Every plant is interesting in its way. A good pigweed is much more satisfactory than a poor rosebush. The pupil should grow the plant from the beginning. He should not buy it ready grown, for then it is not his, even though he own it.

It is well to begin with some plant that grows quickly and matures early. One is ambitious in spring, but his enthusiasm may wither and die in the burning days of summer. If possible, grow the plant in the free open ground; if this is not feasible, grow it in a pot or box or tin can. Take advantage of the early spring enthusiasm. Choose hardy and vigorous plants: sow the seeds when the "spirit moves."

If a pupil is interested in kitchen-garden vegetables, recommend lettuce and radish, or a potato. If in flowers, suggest sweet pea, bachelor's button or blue-bottle, annual phlox, candytuft, China aster. If in fruits, suggest strawberry.

WE DESIRE TO inaugurate a general movement for the planting of plants. The school ground should be planted. Private yards should be planted. Roadsides should be planted. In some cities and villages there are committees or other organizations whose object it is to encourage the planting of public and private places. Sometimes this organization is connected with the school interest, sometimes with a local horticultural or agricultural society, sometimes with a business men's organization. There should be such a committee in every village and town. We wish that the teachers might help in this work, for they would not only be lending their aid to planting, but also be interesting their pupils in some concrete and useful work, and teaching them the value of public spirit. Arbor Day should be more than a mere ceremonial. It should be a means of awakening interest in definite plans for the adornment of the neighborhood and of directing the attention of the children nature-ward.

Gardening and Its Future

When men and women make their homes they gather to
themselves the services and the sentiments that provide life
with both comfort and flavor. The household gods appear one by
one, memories are suggested in a hundred objects that find place
to reside, accessories accrue as the years pass and the sensibilities
enlarge. So it comes that a home is more than a house, and year by
year it is increasingly expressive.

The home needs to express its place in nature. It is a great day
when this fact is discovered and applied. Home-making is not ex-
clusively an indoor responsibility, even though it is within the re-
strictions of a great city. Autumn and winter, spring and summer,
come to every window. Wind and rain deliver themselves at every
door. Sleep and work are conditioned on the season and the day
even though we try to closet ourselves away from them. The better way is to relate the home affirmatively to weather and to place.

The garden is the medium and the agency that accomplishes this
relationship. It is the outdoor part of the home. Its floor is the
earth, carpeted with a thousand forms and colors. Its roof is the
sky. Its walls are trees and horizons. Its library is a multitude of
stories bound in stems and leaves, and ornamented with blossoms.
Its music is the wind and the songs of birds.

Within this area grown men and women and little children are privileged to work. Here they may partake. Here they may open the earth. They may see it absorb the rain. They may plant a seed or a cutting or a root and then may watch it as it grows. They may see the leaf and stem come forth and take shape. They may discover the bud and the flower, and find the seed like that from which this round of life had started. They may cover the span of a life even before they forget its beginning.

They who have a garden will know the first warmth of spring, for they will be waiting. They will hear the first note of new-coming birds. They will see the first butterfly. They will note the greening of the ground. They shall behold a resurrection.

March and April and May will be realities and not merely fugitive sensations. Every month will have its special duty and its particular reward. June will see one garden, July another, August yet another. The seed-time will pass and the harvest will come. Pods and seeds will take place of flowers. Leaves will ripen and fall. Dry dead stalks will remain, record of the hope of April, the fullness of July, and the reward of October. Winter will come, and we shall know that the plants have prepared for it and that they are fulfilling themselves under the snow. The naturalist, the farmer, the real gardener intimately know the seasons. They do not try to escape them. Every season delivers its own satisfactions.

Rains mean something in the garden. They make the beans grow. Winds take the moisture from vegetation and hasten the vital processes. The alternation of day and night is good for strawberries and dahlias. There is adjustment between length of day and welfare of plants. The gardener soon learns to discover the defects in conditions and corrects them as far as he is able. He or she takes care that one organism does not prosper at too great advantage of another. The gardener is the adjuster of the conditions. He has an important range of mastery. The way in which he employs his power will determine his skill and his satisfaction. A garden is a great stimulator of activity and devotion. I have never known a man or woman to start out to make a poor garden.

OF GARDENS THERE are many kinds. The forms are as many as the gardeners. No activity is more adaptable to one's essential wishes and moods. Some persons want vegetables and make a kitchen-garden; others would have color, blue, yellow, red; others desire fragrance; still others wish fruits for the table; others an ever-green garden good in all seasons. Some gardeners incline to the making of designs and others to colonizing plants in borders and woods. Other growers prefer a particular group or class of plants as iris, larkspurs, raspberries, sweet herbs of pleasant memories, lilacs, roses, petunias, lilies, becoming expert and authoritative in them. Some gardeners aim to express their regions or situations and make desert gardens, rock gardens, prairie gardens, bog gardens, alpine gardens. Others grow specimen plants of superior excellence mostly under glass with great explicit skill that we all so much admire and so seldom attain and who need no advice from me. Outside all these amateur ranges is commercial gardening, requiring other abilities. Some persons garden for occupation, some for health, others for release and recreation, the greater number for the joy of it. Other men and women are primarily following the home-making instinct when developing the garden, feeling that without lawn and trees and herbs and the sensation of living with growing things one misses some of the choicest and most enduring privileges. My own gardening is largely of another type. Every year for sixty or more recurring seasons I have had a garden, under my own hand. I have grown all sorts and kinds of plants good, bad and otherwise, outdoors, indoors, often hundreds of species at once, under conditions proper and indifferent. This year I expect to grow a different set of plants. All this may not develop special gardening skill or material for floral display or make a good-looking space, but it ever increases my range of acquaintanceship.

The kinds of plants that one may grow in any climate are legion. The species run into the thousands, and garden varieties of them may be hundreds more. If the horticultural garden plants cannot thrive, there will be native plants to assemble and to grow. The handling of these natives may not be detailed in books, and therefore the effort to grow them should be specially stimulating. No home need lack a garden of some kind if only there is land and sun.

A garden is at first a hope, then a plan, then good personal application. One must give time to it and not be disinclined to read and study; only in that case can it become part of oneself. The prime essential is love of plants rather than admiration of blossoms and colors. Something should be known of the particular soil, of fertilizers, of means of defense against diseases and pests; one should feel the happiness of good tools properly kept, of neat and orderly accessories, and of seed packets labelled and housed. Good records are important, comprising dates of sowing and planting, and the exact names the plants bear. These names and what they signify soon become part of common conversation, and one may then go to bulletins and periodicals and books with increasing understanding and satisfaction.

Gardening is singularly adapted to the education of children, particularly to those who have no free and natural contact with nature. In the garden that is suited to them they may see the processes of life take shape and be partners in the creation. They may find joy in working with their hands and acquire useful practical training; and at the same time they may learn something of the environment in which our lives are set, and I hope may attain an attitude of reverence. The school-garden is one of the most hopeful of educational enlargements, particularly if it is made an integral part of the school process. As we discover ways and means to disperse our population and to give it setting, we should make a garden a part of every school property, and the teaching therein may fortunately be natural, simple and direct.

It is but a step from the garden to the field. There one not only sees plants interesting in themselves but discovers their relationships one with another. Not all kinds of plants grow together indiscriminately. The species differ in adaptations. One kind is associated with another kind of similar requirements, but not with others. One learns what kinds of plants go with clover and what other kinds with mullein and with bullrushes. They have relationships to soils and sunlight and moisture; and there is always the progression from spring till autumn and to spring again. If one's own garden fails, the meadows, roadsides, swales and woods may requite one for the loss. Some of the best gardeners I have known have begun with the fields.

I HAVE BEEN asked to say something about the future of gardening. The future of it will be conditioned, then as now, primarily on the nature of the home. We hope that the horizon of the home is expanding. With bettered physical conditions, better music and books, a more essential education, the fortunate folk of the future will also want better gardens. We are coming into a greater leisure and a vast problem of the future is to learn how to utilize it: gardening should provide one of the major means. Houses will be built increasingly with special reference to well-conditioned gardens. The old garden patterns and perhaps the old plants will not satisfy. I apprehend that the plant lovers of the future will prize species or original stocks of plants that yet mean little to us, if indeed we have even grown them. The earth has not yet been sufficiently explored for plants. Perhaps discriminating persons in the centuries to come may care relatively less for the confused products of hybridization. With better plant materials and more sensitive souls, great changes may come in the development of parks and public places. Crude refuse wastes will disappear. Whole countrysides will come under the domain of the artist, and gardening, in its expanded sense, will become co-extensive with man's occupancy of the earth. Tastes in plant-growing may be expected to change as much as taste in literature and art. Science will demand new reactions. The people in that good day may look back on the best materials and the noblest books of our present moment as amusing curiosities. It is our part to make this consummation possible; and with this pleasing prospect I leave you to your reflections.

Undertone

From morning till night and everywhere
My days are full of their effort and care;
Full of labors to drive and schemes to test,
Of work to finish and knowledge to wrest;
And the known result of this noise and strife
Is what men and the world all call my life,—
This is the meed of the work that I own
Outspread on my life as an overtone.

But ever there runs through the work I own
The all-silent stream of an undertone.
This stream is myself as my life I live
And out of it flows all the strength I give.
It's the tone of hills and calm of the plain
The smell of the soil and the touch of rain;
'Tis a careful thought of the calm sweet grass
An abiding joy in the birds that pass
In the mite that lives in the growing shoot
And the changing tints of the leaf and fruit;
'Tis the melting snows and the morning sun
And the soft gray days and the marshes dun;
'Tis appeal of frost and the fragile dew
Of the passing clouds and the depths of blue;—
Then a quiet heart that can give no sign

Of the sacred calms that are only mine,
Or the gentle sins that are part of me
As the silent twigs are part of the tree,
Or memories deep I cannot express
Any more than the tree in its wild'rness.

The peace of the winds is my undertone—
I move with the crowd, but I live alone.

II

THE GROWING OF THE PLANTS

As there must be many gardeners, so there must be many books. There must be books for different persons and different ideals. The garden made by one's own hands is always the best garden, because it is a part of oneself. A garden made by another may interest, but it is another person's individuality. A poor garden of one's own is better than a good garden in which one may not dig. Many a poor soul has more help in a plant in the window than another has in a plantation made by a gardener.

The Miracle

A friend gave me a particle as dry and brown as a grain of sand. He said it was a seed. He told me that if I would put it in the earth and then watch the place, I should behold a miracle.

Presently a tiny thing appeared, green, with two leaves. My friend declared it came from the particle I had buried. It did not seem possible. It had no mark or semblance of the globular wrinkled seed; and how had the seed found itself among all those grains of sand?

Upward it grew, adding leaf on leaf, all unlike those that first I saw. Hairy stems struck out here and there. Buds came, unlike the leaves or the stems. Then came flowers in gorgeous color, unlike the leaves in color or form or substance.

There it grew, this miracle from the seed. The earth in the pot was black-brown and formless. The water that I added was color-less and formless. The air in which it grew was invisible. Yet here were upstanding brown-green stems, fragile but strong; wide expanding leaves of green, with scalloped edges, thin and veiny, soft with velvet to the touch; crimson flowers on long stems, more fragrant than the dew, shapely, bearing delicate organs within; and presently there came pods, and in them I saw seeds like the one I had planted. What a succession was here of objects all unlike each other, all coming out of the same earth, the same water, the same air!

A seed that my friend planted made stems and leaves and flowers and pods as different from mine as a cat is different from a dog. I wondered how these things could be. I have asked many wise persons, but none of them can tell me.

How to Make a Garden—
The First Lesson

It is customary to begin with the preparation of the soil when giving advice on the making of gardens. The real beginning of a garden lies farther back, however. It is a mental concept. One does not begin to make a garden until he wants a garden. To want a garden is to be interested in plants, in the winds and rains, in birds and insects, in the warm-smelling earth.

The best preparation for garden making is to go afield, and to see the things that grow there. Take photographs in order to focus your attention on specific objects, to concentrate your observation, to train your artistic sense. An ardent admirer of nature once told me that he never knew nature until he purchased a camera. If you have a camera, stop taking pictures of your friends and the making of mere souvenirs, and try the photographing of plants and animals and small landscapes. Notice that the ground glass of your camera concentrates and limits your landscape. The border-pieces frame it. Always see how your picture looks on the ground glass before you make the exposure. Move your camera until you have an artistic composition,—one that will have a pictorial or picturesque character. Avoid snap-shots for such work as this. Take your time. At the end of a year, tell me if you are not a nature-lover. If to-day you care for only pinks and roses and other prim garden flowers, next year you will admire also the weedy tangles, the spray of wild convolvulus on the old fence, the winter stalks of the sun-flower, the dripping water-trough by the roadside, the abandoned bird's nest, and the pose of the grasshopper.

What has all this to do with gardening? Where I write is a child's flower garden. In one spot the flowers failed. Here a bushy fire-weed has filled the space. The child has let it grow, and it is part and parcel of the garden. The child knows that a fireweed is better than hot bare earth, and, fortunately, no one has ever told her that a fireweed is not beautiful. Her contact with nature is one point more intimate than it would have been without the fireweed, and she will always feel that each plant is interesting after its kind. The fireweed has no showy flowers: she likes it because it is a plant and fills a place in her garden, not because it fills the eye with color.

Appreciation of the charm of plant form and color is the real beginning of gardening; and this is why gardening appeals with such power to persons of refined tastes. Every plant is interesting. Yet the person is discriminating, loving one plant more than an-other or with a different kind of love. There are times and seasons for all plants. One's sympathies are wide, as one's life is full and resourceful.

In its largest sense, the garden comprises the personal parts of the estate,—those parts that are not devoted to specific agricul-tural purposes. The lawn, the grove, the park, the flower garden, the vegetable garden, the fruit garden for the home supply,—all these partake of the garden spirit rather than the farm spirit, and these are intended in the sketch I am writing. The beauty of a tree, therefore, is as much a garden subject as the beauty of a flower. I like to think that there are four epochs in the life of any garden plant, four kinds of interest to the cultivator. These are, first, the juvenile stage, when the initiate is coming into its first contact with the problems of living and when its future is all to be determined; second, the aspiring growing stage, when every day sees change and progress; third, the blooming and fruiting stage, the only one of the four that appeals to most people; and fourth, the stage when the work is done and the leafless framework stands bare above the brown earth or is buried in the drifting snow. To me, the first and the fourth are the most significant and the most mysterious. I am willing to take the trouble of growing a plant that I may see its leafless stems above the snow on Christmas day.

Not in flowers alone do plants appeal to the true gardener. Not long ago I was interested in observing the different points of view of such a gardener and his customer. Born in a foreign land and widely traveled, the gardener had always loved plants because they were plants. He was now growing plants for a living on our Pacific coast, but his establishment had none of the rich and florid look of the commercial garden. Plants of mean habit and small flower, plants common-looking and inconspicuous, tucked away in nooks and corners—these he pointed out to her with almost pathetic appeal as she led him on and on in quest of some gaudy coleus or geranium. The latter she purchased, in spite of his earnest recommendation of the humbler and really prettier things; but I thought that he was glad that she spared the ones that were nearest his heart.

It is good practice for the beginner consciously to compare the general habits and forms of plants. Contrast any two. Learn their characteristics, the size, the direction of growth, the general swing of the branches, the hang of the spray, the amount of foliage and its color and texture, the mass effect of its flowers. Make out what the framework or scaffolding would be if all the leaves and flowers were removed. Compare, for example, the compact habit, drooping spray and garland-like inflorescence of the mock-orange, with the open habit, and horizontal airy spray of the Japanese snow-ball; or contrast the climbing withes of clematis with the bole from which it hangs. Note how these plants change from spring till fall, and from fall till the quickening days of spring. Believe that every plant lasts at least a twelvemonth.

The bolder parts of the garden should be taken by the bolder and stronger plants. The humbler and transient and more exotic things should be the incidents. Throw in the borders thickly and profusely. Plant only sparingly in the centers. Leave some generous space of open greensward as a contrast and a foreground for the planting. The most dangerous thing in the planting of a place is to encroach on the open lawn, for one plant out of place there may introduce an element so discordant as to spoil the whole effect of the picture. As one would embellish a building with minor architectural effects and tasteful combination of colors, so the gardener can brighten and enliven the premises by means of every kind of choice and cheery flower that will bloom in the nooks and along the borders. In the distance may be iris, zinnia, marigold, dahlia, sunflower, foxglove, hollyhock, each in its place and of sufficient quantity to produce a forceful local impression of individual habit and color. In the foreground may be pinks, bachelor's button, the annual phlox, verbena, deep-tubed petunia and all the minor citizens of the garden. Make up your mind in advance that you will like all the plants that grow in your garden—then your gardening will be successful.

The Home Garden

Simple desires, with every well planning and well carried out, result in the best gardens.

The garden must be yours; if it is another's it is not worth the while to you.

A good garden is the one that gives its owner the most pleasure; he may grow orchids or thistles.

The measure of success in the garden is the sensitive mind rather than the plants.

The home garden is for the affections. It is for quality. Its size is wholly immaterial if only it have the best. I do not mean the rarest or the costliest, but the best—the best geranium or the best lilac. Even the fruit garden and the vegetable garden are also for the affections: one can buy ordinary fruits and vegetables—it never pays to grow them in the home garden. When you want something superior, you must grow it, or else buy it at an advanced price directly from someone who grows for quality and not for quantity. If you want the very choicest and the most personal products, almost necessarily you must grow them: the value of these things cannot be measured in money. The commercial gardener may grow what the market wants, and the market wants chiefly what is cheap and good looking. The home gardener should grow what the market cannot supply, else the home garden is not worth the while.

A garden is a place in which plants are grown, and "plants" are herbs and vines and bushes and trees and grass. Too often do persons think that only formal and pretentious places are gardens. But an open lawn about the house may be a garden; so may a row of hollyhocks along the wall or an arrangement of plants in the greenhouse. Usually there is some central feature to a garden, a theme to which all other parts relate. This may be a walk or a summer-house or a sun-dial or a garden bed or the residence itself, or a brook falling down the sward between trees and bushes and clumpy growths. There are as many forms and kinds of gardens as there are persons who have gardens; and this is one reason why the garden appeals to every one, and why it may become the expression of personality. You need follow no man's plan. The simplest garden is likely to be the best, merely because it is the expression of a simple and teachable life.

Grow the plants that you want, but do not want too many. Most persons when they make a garden order a quantity of labels. Fatal mistake! Labels are for collections of plants—collections so big that you cannot remember, and when you cannot remember you lose the intimacy, and when you lose the intimacy you lose the essence of the garden. Choose a few plants for the main plantings. These must be hardy, vigorous, sure to thrive whether it rains or shines. These plants you can buy in quantity and in large, strong specimens. Each clump or group or border may be dominated by one kind of plant—foxgloves, hollyhocks, spireas, asters. The odd and unusual things you may grow as incidents, as jewelry is an incident to good dress. Miscellaneous mixtures are rarely satisfactory. The point is that the character of the home garden should be given by the plants that are most sure to thrive. The novelties and oddities should be subjects of experiment: if they fail, the garden still remains.

The lawn should be the first care in any home ground. All effective planting has relation to this foundation. Homelikeness also depends upon it. Grass will grow anywhere, to be sure, but mere grass does not make a lawn. You must have a sod; and this sod must grow better every year. This means good and deep preparation of the land in the beginning, rich soil, fertilising each year, re-sowing and mending where the sod becomes thin. Usually we water our lawns too much, making the grass shallow-rooted and causing it to fail early. Every inducement should be made for the grass roots to go down.

In very shady places, as under trees and wide eaves, it is very difficult to secure a good sod. In such cases we must rely on other plants for the carpet-cover. Of these other plants, the best for the North is the common running myrtle, or periwinkle. Sods of this make an immediate and persistent cover. Lily-of-the-valley also makes a fairly satisfactory groundcover in some places. If the soil is damp, the moneywort may be tried, although it sometimes becomes a pest. Take note of the ground-cover in all shady places that you come across. You will get suggestions.

Put walks where they are needed—this is the universal rule; but be sure they are needed. In the beginning you will think you need more than you actually do need. How to get the proper curve? Perhaps you do not need a curve. There are two fixed points in every walk—the beginning and the ending. Some walks lack either one or the other of these points, and I have seen some that seemed to lack both. Go from one point to the other in the easiest and simplest way possible. If you can throw in a gentle curve, you may enhance the charm of it; and you may not. Directness and convenience should never be sacrificed for mere looks—for "looks" has no reason for being unless it is related to something.

For main walks that are much used, cement and stone flagging are good materials, because they are durable, and they keep down the weeds. There is no trouble in making a durable cement or "artificial stone" walk in the northern climates if the underdrainage is good and the cement is "rich." For informal walks, the natural loam may be good; or sharp gravel that will pack; or cinders; or tan-bark. For very narrow walks or trails in the back yard I like to sink a ten-inch-wide plank to the level of the sod. It marks the direction, allows you dry passage, the lawn-mower passes over it, and it will last for several years with no care whatever. In flower gardens, a strip of sod may be left as a walk; but the disadvantage of it is that it retains dews and the water of rainfall too long. Some of the most delightful periods for viewing the garden are the early morning and the "clearing spell" after a shower.

There should be no fence unless there is a reason for it. Some persons seem to want fences just for the purpose of having them. Of themselves, open fences are rarely ornamental or desirable. They are expensive property. The money put into a fence will often buy enough plants to stock the place. Front fences, in particular, are rarely desirable. The street and the walk sufficiently define the place. Now and then a person wants a front fence to give his place privacy. This may be a perfectly legitimate desire, but the requirements are usually best satisfied by means of a low and substantial wall. A fence means protection. A wall may mean seclusion; and it may easily be made a part of the architectural features of the place. Walls usually work well into the planting designs of a home ground, but the instances where fences do this are exceedingly rare.

Even in the back yard a wall may be preferable to a fence, but pecuniary considerations may determine for a fence; and, moreover, a real fence is more in keeping in a rear yard, for that yard is usually most in danger of molestation. In the back yard, the fence may become also a screen and a shelter. Usually it can be covered with vines—sometimes with grapevines—to advantage, or be "planted out" with bushes and trees. It is good practice to allow the fence to obtrude itself as little as possible.

As a whole, the garden is maintained for its general effect. It is a part of an establishment, of which the residence, the barn, and the boundaries are other parts. But the garden should also have certain parts that are for distinct or particular service, that should be to the general garden what pantries and bedrooms and closets are to the house. These garden-rooms are for vegetables or flowers or fruits or sweet herbs. These things are grown for use in the family, not for their effect as a part of a garden picture. They can be grown best in special areas set aside for this particular purpose, where the soil can be regularly tilled and each plant given full room and conditions to develop to its best. This is as true of flowers as it is of beets or strawberries. The fact that we grow flowers also as part of the garden picture should not obscure the fact that we also grow them for cutting and for decoration and exhibition. When China asters are wanted because they are China asters, grow them where and how China asters thrive best; if they are wanted as a part of the general garden effect, grow them where and how this effect can be best secured.

The place for the service garden is at one side or the rear—preferably in the back yard. Grow the things in rows.

Give the children an opportunity to make a garden. Let them grow what they will. Let them experiment. It matters less that they produce good plants than that they try for themselves. A place should be reserved. Let it be well out of sight, for the results may not be ornamental. However, take care that the conditions are good for the growing of plants—good soil, plenty of sun, freedom from the encroachments of tree-roots and from molestation of carriage-drive or chickens. It may be well to set the area off by a high fence of chicken-wire screen; then cover the fence with vines. Put a seat in the enclosure. This will constitute an outdoor nursery room; and while the child is being entertained and is gaining health he may gain experience and nature-sympathy at the same time.

There are two kinds of interest in plants—the interest in the plant itself for its own sake, and the interest in plants as part of a mass, or as elements in a picture. The former is primarily the interest of the plant lover or the botanist; the latter is the interest of the artist. Fortunately, many persons have both these elements highly developed, and every person can train himself to appreciate both points of view. Now, a home ground is one thing. It is, or should be, homogeneous in its composition. It should appeal to one as a unit: the entire place should produce one effect. This effect may be that of rest or retreat or seclusion or homelikeness. In order to produce this harmony, plants must be placed with relation to each other and to the general design of the place. The ability to do this kind of planting is one of the attributes of a good landscape gardener. He produces good "effects" and harmonies. He thinks less of plants as mere plants than he does as parts of a composition. He sees them much as a painter does. All this is contrary to the general conception of planting. Most persons, I fear, think of a plant only as a plant, and are content when it is planted. But merely to plant a plant may have little merit in the home grounds: robins and squirrels do that much.

How to Make a Garden— Digging in the Dirt

With the first break of spring, digging in the dirt is pleasure. When digging becomes a necessity in order to cause plants to grow, it is labor.

We enjoy the doing of things that other people cannot do; also of those things that have some special personal application to ourselves. Digging anybody can do. It is an exercise of mere muscular effort.

If this is your idea of digging and plowing, you will have little pleasure in making a garden, for much digging is necessary. When you really come to know what digging in the garden means, and what forces it sets at work, you will enjoy it—up to the limit of your physical powers—because it will be intellectual occupation. You will learn to economize muscular effort when you once understand the philosophy of tilling the land. Blind work in the garden has driven many a boy from the farm. Do you remember the fenced-in little garden, into which a horse could not enter, with onions and parsnips and "pusley" growing in hostile confusion on those pretty little beds lying between narrow sunken walks? Yes; well, they are the gardens I mean. Those beds always needed weeding, and the weeds were thickest and toughest-rooted on the Fourth of July, circus days, holidays and the blistering afternoons of summer. There was no idea of tillage; it was all weedage. More effort was expended merely in fingering out the weeds than all the turnips and carrots were worth; and as soon as the weeding made the tender things visible, the bugs laid quit-claim on the whole bed. If those gardens had been laid out in long, comfortable rows so as to allow of the use of the wheel hoe—if not of the horse cultivator—there could have been efficient stirring of the soil as well as the mere killing of weeds; and then there might have been some fun in raising beets and parsley. But there was discipline for weeds and boys in the old checker-board garden!

Why We Till

To till the land is to stir it to the end that plants may thrive. In our loose colloquialism we speak of this practice as cultivating; but cultivation comprises more than tillage. We till the land for the land's sake, not primarily to kill weeds. In the best-tilled lands, weeds have small opportunity to grow. Land that is wholly clean of weeds may need tillage quite as much as that which supports an usurious growth of them.

We till—

1. To improve the texture or physical condition of the soil;
2. To hold and save the moisture;
3. To promote and hasten chemical activities, and to make it possible for the countless micro-organisms to live and thrive;
4. To kill plants that we do not want—weeds;
5. To enable us to get the seed into the land and in some instances to get the crop out of it. These two last are sometimes considered to comprise the whole art and purpose of tillage. In such cases stirring the land is interesting only when it is of short duration.

Texture is as important to good soil as it is to a piece of dress goods. The table on which I write contains an abundance of plant-food, but the table will not grow plants: it lacks the proper texture and the moisture. The adjectives that farmers and gardeners apply to good and poor land are suggestive of texture rather than of richness in plant-food, as "mellow," "mealy," "friable," "loose," "crumbly," "warm," "quick," "cloddy," "hard," "dead." Even when one says that a soil "looks rich" he does not refer primarily to its plant-food content; by experience he has learned that a soil that "looks" mellow, soft and dark-colored is capable of growing good plants. The superlative benefits of stable manures usually arise more from their ability to improve the texture of the soil than to add to its richness in chemical plant-foods.

Without water, plant-growing is impossible. In most parts of the older settled regions, there is sufficient rainfall to maintain good crops. But this water must be saved. The soil must be put in proper condition to receive and hold it. The soil should be loose and mellow, for such soil has capillary power and thereby holds much moisture. The greater the depth to which this fineness extends, the greater the amount of moisture that can be held. If the soil contains a liberal supply of humus—secured from vegetable matter, as stable manure and green-crops and litter turned under—it has greater moisture-holding capacity. If lands are hard and bare when winter comes, much of the precipitation flows away as surface wash. If the soil is newly plowed or spaded in the fall, the rain and snow are held and surface drainage is lessened. If the soil contains a cover of vegetation, much of the rainfall is held until it can soak away. But the amount that is held depends, as indicated above, by the mellowness and depth of the soil. The water-table is the region of standing or free water in the soil. It is the bottom of our dish-pan. The lower the water-table the greater the capacity to hold capillary or hygroscopic water,—and this water is that which is used by most agricultural plants. The water-table is lowered by deep tillage, underdrainage, the growing of deep-rooted plants. The deeper the water-table the less likely is the land to become muddy—for mud occurs when the dish-pan is full or runs over.

Once having trapped the water, we must keep it. We do this by lessening evaporation from the surface. Spread a mulch over the land. The cheapest mulch is loose, comparatively dry earth, and this mulch can be secured by frequent shallow stirring of the surface soil. This is the prime value and philosophy of surface tillage,—to make and keep a loose earth-cover. The soil never should "bake." When this mulch is well maintained—by rake or harrow—weeds have no opportunity to grow. Weeds advertise the faults of the farmer.

How We Till

We till (1) to prepare the land, (2) to maintain its condition. On hard lands, preparation-tillage usually should be deep in order to deepen the water reservoir and to increase the foraging area of the roots. On loose and sandy lands with no evident water-table, deep plowing or deep spading may be unnecessary unless it is desired to work manure deep into the soil. Some persons have taught that deep plowing is a principle, but it is only a practice.

But whatever the depth of preparation-tillage, it should be thorough. Work up the soil carefully; then go over it again; then again. Good preparation is worth more than fertilizers. In fact, fertilizers are valuable only when the soil is in such good condition that the plants have a chance of thriving. Till first, and add plant-food afterwards.

If the land was not broken last fall, spade it or plow it as early as possible this spring, in order to prepare it to hold the spring rainfall. Put on the surface mulch with the rake or harrow, to keep the moisture in the soil. Even if the crop is not to be planted until June, the land should have surface tillage from the time that it is dry enough to work. Hold the moisture against the time that it is needed. Many persons fail with the crop because they lost so much moisture before the crop was planted that there was not sufficient left to carry the plants to maturity. In the early season, the "spring fever" is on. We really want to dig. Take advantage of your own ambition: do the hardest labor first; merely keep the land in condition during the dog-days.

The novice should be cautioned not to stir the land when it is wet, particularly if it contains much clay or silt. When land is dry enough to crumble, it may be worked. Stirred when too wet, it may become pasty and lumpy; the texture may be injured rather than improved.

Persons often ask what is the best tool for tilling the surface of the land—to make the earth-mulch. The best tool is the one that does the work best and expends the least amount of energy in doing it. On hard soils, a spading-harrow or disc-harrow may be the best tool for the first tilling after the plowing is done. A spring-tooth harrow may then follow. For the final fitting, a smoothing-harrow or a weeder may be best. The point is that the kind of tool must be selected with reference to the soil and the result to be accomplished.

In gardens, save the moisture by means of frequent surface raking between the plants.

Every week, at any rate, it is well to rake the surface thoroughly, even though the soil is dry and weedless. After a rain always stir the soil before it bakes. Do not carry water in pails or run it through a hose, until you have saved all you can of that which falls from the clouds. Water your garden with a rake.

Perhaps these random remarks will open a new world of interest to you. I hope you may some day understand the subject so well that you will till the soil for the fun of tilling it, not merely because tilling is necessary nor even because it affords mental and physical relaxation. You will follow the plow with a new sensation. The crumbling furrow will suggest a thousand problems. You will hear the plants laugh. The soil will have a new meaning to you. Luther Burbank wrote me the other day that I should see his garden with "the asparagus punching up the mellow earth with joy."

"The asparagus punching up the mellow earth with joy."

The Growing of Plants by Children—The School-Garden

A ctually to grow a plant is to come into intimate contact with a specific bit of nature. The numbers of plants that we grow, and also the kinds of them, increase with every generation. The intensity of our plant-growing, as well as the increasing care for animals, is coming to be a measure of our interest in the world about us.

Not only has the cultivation of plants itself increased our contact with plants and with nature, but, in connection with the growth of the spirit of art, of sport, and of suburbanism, it has taken us afield and has impelled us to know things as they are and as they grow. The modern popularization of plant-knowledge is probably due more to these agencies than to the progress of botany.

There are many practical applications to the lives of children and to the home that may be made from a knowledge of plants and horticulture. This knowledge means more than mere information of plants themselves. It takes one into the open air. It enlarges his horizon. It brings him into contact with living things. It increases his hold on life. All these facts were well understood by Froebel, Pestalozzi, and other educational reformers.

It is important that one does not assume too much when beginning plant-work with children. We forget that things which fail to appeal to us, because of our busy lives and great experience, may nevertheless mean very much to the child. Often we attempt to teach the child so much that it is confused and nothing makes an impression. An interest in one simple, living problem that is near to the child's life is worth a whole book of facts about nature.

It is not primarily important that children know the names, although the name is an introduction to a plant as it is to a person. The essential point is that there should be plants about the home, or in the school grounds, or in the schoolhouse windows. Even though the children are not conscious that they are receiving any impression from these plants, nevertheless the very presence of them has an influence that will be felt in later life, even as the presence of good literature and furniture and the association of refined surroundings has influence.

I dropped a seed into the earth. It grew, and the plant was mine.

It was a wonderful thing, this plant of mine. I did not know its name, and the plant did not bloom. All I know is that I planted something apparently as lifeless as a grain of sand and there came forth a green and living thing unlike the seed, unlike the soil in which it stood, unlike the air into which it grew. No one could tell me why it grew, nor how. It had secrets all its own, secrets that baffle the wisest men; yet this plant was my friend. It faded when I withheld the light, it wilted when I neglected to give it water, it flourished when I supplied its simple needs. One week I went away on a vacation, and when I returned the plant was dead; and I missed it.

Although my little plant had died so soon, it had taught me a lesson; and the lesson is that it is worthwhile to have a plant.

Provide some little means of growing plants, not only to teach how to grow plants themselves, but to teach the child the care of things, to show that other beings besides itself have vicissitudes and lives of their own, and to implant the germ of altruism—the interest in something outside of oneself. These means of growing plants should be simple. A pot, a box or a hotbed may be sufficient. Every child should have the handling of at least one plant during the period of childhood. One plant cannot be handled without leaving an impression on the life.

The love of plants should be inculcated in the school. It can usually be better done in school than at home, particularly when one or both of the parents is opposed to it and constantly discourages the child. Even when the parents are ready and competent, the teacher may be able to reach the children more effectively than they. In nearly every school it is possible to have a few plants in the window. They may not thrive, but it is worth while to set the children to inquiring why they do not. Sometimes the poorest plants awaken the most effort and inquiry. If nothing else will thrive, a beet will. Secure a good fresh beet-root from the cellar. Plant it in a box or tin can. Surprisingly quick it will throw out clean bright leaves. The thick root will hold moisture from Friday to Monday.

A desire for school-gardens is gradually taking shape. This movement must grow and ripen; it cannot be perfected in a day. Through the centuries there have been few school-gardens: we must not expect to overcome the lack at once. The movement has not been aided much, if at all, by those who have "complete" schemes for gardens for the district schools. Such schemes may be advisable later. Start the work by suggesting that the school-grounds be cleaned or "slicked up." Take one step at a time. The propaganda for school-gardens must have relation to the economic and social conditions under which the school exists.

There is some confusion as to the objects of school-ground improvement. The purposes may be analyzed as follows:

1. Ornamenting the grounds, comprising (*a*) cleaning and tidying them, (*b*) securing a lawn, (*c*) planting. This is always the

first thing to be done. It stands for thrift, cleanliness, comfort, beauty, progressiveness.

2. Establishing a collection to supply material for nature-study and class work.
3. Making a garden for the purpose of (*a*) supplying material (as in No. 2), (*b*) affording manual-training, object lesson work, and instruction in plant-growing.
4. Providing a test ground or experiment garden where new varieties may be tried, fertilizer and spraying experiments conducted, and other definite studies undertaken.

These purposes fall into two main groups: (1) The improvement or adornment of the grounds; (2) the making of distinct gardens for purposes of direct instruction, or school-gardening proper. Much of the current discussion does not distinguish these two ideals, and thereby arises some of the loss of effort and effectiveness in the movement.

The plat on which the three hundred and seven kinds of plants were grown this year measures about twenty by forty feet. It is an ordinary piece of land such as one finds in the rear of a residence, partly made by filling, and put into condition year after year by timely tillage, abundant humus and more or less chemical fertilizer. As the plat is in no way unusual, neither were the plants unusual except as to their kinds.

At the outset the reader wants to know why any one in his senses would want to grow so many kinds of plants. Presumably the tendency is to cut down the number of kinds; if so, then I assume that folks wish to reduce their chances of enjoyment. Be that as it may, the three hundred and seven were planted on this little patch because there was no other place to grow them. The remainder of the garden was taken by a still greater number of kinds; and seeds and roots accumulate rapidly if one so desires.

The larger part of the three hundred and seven were distinct species of plants. Only a small number represented horticultural varieties, and even these were mostly under Latin names as if they were species. There were no florist's plants among them, none that would now attract a window customer. Of course, not all the number found perfect conditions on the little plat and perhaps some of them did not produce their best; this was not to be expected; but they were not grown for display or to develop the most perfect specimen plants; nor was anyone invited to see the garden, for it was not of the usual kind and not in competition and perhaps the visitor would not think it a garden at all. Yet I felt it the best annual garden I ever made. To make amends the area will go into a plain flower-garden next year, with colors to fill the eye. Last year it was a vegetable-garden.

Why did I grow them? Because I wanted to find out what they were, and there was no other way; and if there had been any other way I should have grown them just the same, not caring to miss such a good opportunity. These plants were offered by seedsmen and others; it is good to know what plants are in cultivation. My reader, wishing a reason, may call it a specimen-garden.

No, THE PLANTS were not much crowded, at least not for my purpose. There were twelve rows running lengthwise the plat. The reader can calculate the distances, and the length of row given to each kind; I have neither the time or inclination. I put them all in myself, on my knees; this develops the proper mental attitude. The little row or drill is made along a garden-line with the point of a triangular hoe. All the seeds in a row are sown at one operation, in the bottom of the open drill, choosing for each row plants of similar size if other circumstances allow. Stake-labels have previously been written and these are securely placed before the seeds are dropped. Then the seeds in the entire row are covered by sifted well-rotted mold obtained from my perpetual compost of leaves and litter, with no soil on top; then the mold is pressed down by the blade of a stout hoe, my foot on the blade.

How they grew! Sown the first week in May they came up in legions, for I always sow thickly to insure a stand, to allow for weak seeds, and to obtain the pushing effect of a lot of sturdy shoots; one can afford this liberal seeding in two feet or less of row. The thinning was begun almost as soon as the plants appeared, for one cannot correct the injury of overcrowding in these tender things. With a flat label or a small hand-weeder, and with fingers, great numbers of the plants are removed at once. Before any injury was wrought, the plants were thinned again, and perhaps yet again and again. About two or three good plants of each kind were wanted; but to allow for accidents, four or five or more plants matured. Some of them are staked if need be. Neighboring plants are restrained. The rows are gone over every day or two. Every first bloom is seen.

What "effect" was obtained from this garden? the reader is asking. None. I was not working for effect, not even for bloom in the usual sense, but for plants. Of course the result lies in the record; and the record is a good specimen (perhaps several specimens) on an herbarium sheet, with the dates, the color, and any other note not shown by the specimen itself; and the seed-packet is with the specimen: the reader may learn more about this form of recreation if he has the patience to read the article called "Candytuft."[1] Somewhere there should be a record of the things in cultivation. It will be needed some time.

No, this form of gardening is not recommended to anyone. It is reserved specially for my own delight, and I have been addicted to it for so many years that I cannot forego it now as the day approaches the twilight. The plants in my garden and in all wild free places have been my companions. How many are the thousands and the thousands that have complimented me by growing in my garden, and what memories they release!

Choice associations these plants bring to me. Amongst the three hundred and seven were twenty-six kinds of lupines: species

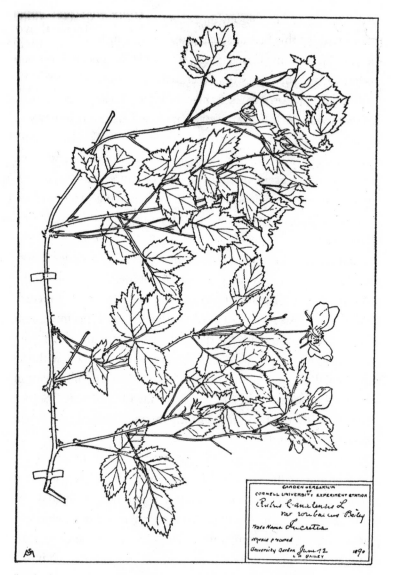

GARDEN HERBARIUM
CORNELL UNIVERSITY EXPERIMENT STATION

Rubus Canadensis L.
var *roribaccus Bailey*

Local Name *Lucretia*

Whence procured

University Garden June 12 1890
L. H. BAILEY

"*An herbarium sheet of a form (var. roribaccus) of* Rubus villosus."

yellow, purple and violet, blue, white, big-seeded and little-seeded, smooth and hairy, short and small and tall and big, native in Mexico and Guatemala, southern Europe, California, South America, Oregon, Texas, and otherwheres, all growing side by side where not one of them is naturally at home. Is it not wonderful?

Surely I should be thankful to collectors and seedsmen who have assembled these treasures and have made it possible for me to have them almost without effort in my little garden, all by myself. Here following each other in a row are two species of Collinsia (commemorating Zaccheus Collins, an early leader in Philadelphia) from California, the odd fleshy compositous Cryptostemma from South Africa and Australia, delicate Diascia from South Africa, Dimorphothecas waiting for the sun, an Elsholtzia from Asia (not to be confounded with the very different Eschscholzias of California), brilliant Eutocas from California that may be named in Phacelia if you prefer, several pleasing Gilias from the West, little western Leptosiphons that may be placed in Gilia or Linanthes,[2] *Grahamia aromatica* "in pastures and shrubby hills" of Chile and now placed with the Heleniums, the pleasant *Helipterum corymbiflorum* of Australia, Lasthenia from California, indefinite *Leuceria senecioides* of Chile, button-headed Lonas (or Athanasia) of the Mediterranean country, brilliant Tagetes from Mexico, a dainty red-flowered little Spergularia not yet made out, and others.[3] What a goodly internationalism is this! And what is the magic that enables all these diverse things from around the wide world to thrive in the one soil in my obscure garden? The world is surely a vast and pleasant democracy.

THE READER WILL note that some of these plants have two or three names and this may affright those people who demand an invariable nomenclature. Botanical nomenclature is part of a living technical language that has been evolving for hundreds of years. It can never be made invariable, nor ought it to be petrified into rigidity.

Some of the best history of plants is suggested in the Latin names; and there may never be agreement as to whether Eutoca should be referred to Phacelia, whether Grahamia as a genus is really distinct from Helenium, for there is no finality in natural objects and no more possibility of agreement among botanists than among politicians, churchmen, grammarians, economists, horticulturists, or others on the subjects with which they deal, and, moreover, the naming of plants follows recognized rules of procedure in scientific work; in the natural history of the case, stability in names is unattainable. The different interpretations are of real value in the record; and the more synonyms a plant has the greater is the interest in it, because the many names suggest perhaps wide distribution or great variability or a long devious history in the books, or knowledge of it on the part of many persons. Most of the synonyms are not mere dead names, as so many persons suppose; to the practiced eye they open at once a suggestive field of interest and investigation. To know the plant, one must know something of its history.

It is said that a uniform and constant set of names is needed for purposes of trade. If so, let a competent body of delegates, representing "the trade," adopt an arbitrary list to be used in commercial catalogues for a given series of years; and let the list be revised at the conclusion of the periods, to keep in touch with the important necessary changes. But such an arbitrary list can never constitute a code of nomenclature or take the place of rules to determine cases, nor can it be adopted by working botanists except insofar as it happens to conform to essential botanical practice. Nomenclature must be natural and controlled by law, never despotic. It must be at least as variable as are the plants. Some of the present botanical disagreement is of course unfortunate, and continued effort should be made to unify the practice; but if horticulturists had a better understanding of the principles and precedents underlying the naming of plants, the subject of nomenclature would have few terrors, as has been suggested in an earlier page. The more exact one's knowledge of plants the less perplexing is the nomenclature. Perhaps a thousand names have been applied to my three hundred and seven, but they do not detract from the ease, precision and joy of growing and knowing the plants.

AMONG THE THREE hundred and seven are some fifty kinds of grasses. They are a fascinating lot. They bear no resemblance to the twenty-four kinds of œnotheras alongside, although sown the same time in the same soil and tilled by the same tools in the same hands. I do not know why, except that grasses are grasses and œnotheras (or evening-primroses) are œnotheras; but this is not an explanation. Among themselves these grasses are various enough for any taste, from the hair-like tufts of fescue three or four inches high to the rigid straight culms of a Pennisetum at the end of the row nine to ten feet tall. Both these plants grew in the same number of days. There are dangling Brizas, lawn-like Poas, capillary Panicums, and the heavy stalks of Caragua corn. Many choice ornamental things are among these grasses, fit for mass-planting, edgings and bouquets. Yet in our day the grasses will not be much sought as ornamental subjects because not sufficiently showy. They do not advertize or obtrude themselves and would not consort well with the gaud and emphasis necessary to make such things popular. The fact that so few persons choose the modest plant-forms should make these forms the more attractive to such as desire a more personal expression, or who are interested in minor effects.

It is not profitable to enumerate more of the kinds, although I can hardly refrain from mentioning the mats of Herniaria, the excellent *Amellus annuus*, good first-year Campanulas, graceful *Baeria gracilis*, *Ambrosia mexicana* (which is the persistence of an old herbalist designation for the Jerusalem-oak or *Chenopodium botrys*), renascent Anastatica, bright Nemophilas, a good lot of Nemesias and Linarias, *Viola bosniaca*, Abronia, Anoda, aromatic Gardoquia or Cedronella, white Ismelias, Anacyclus, Spilanthes, blue Nolana, stalwart Pychnostachys,[4] ephemeral Commelina, and showy Calandrinia. Some of these names may seem strange to the American gardener, but they are as interesting for all that.

With this fragmentary enumeration some of my readers will more than ever ask what is the use of it all. Such a miscellaneous lot will by them be cast into the discard of "novelties," which is the sufficient and final condemnation. It is the commercial men who most object to a large number of varieties, and for their purposes they are right. The canners properly reduce their varieties of corn and beans and peas to the very few that meet their needs; the commercial orchardists cannot afford a miscellaneous variety of peaches, apples or grapes; the florist knows only a few profitable roses for the cut-flower trade. I am in complete sympathy with the reduction and standardizing of these varieties for trade purposes. But this is no reason why the commercial men should propose the elimination of variety for other people,—that would be an impertinence.

Let not my reader think that all the three hundred and seven were worth the effort beyond the study and record. A good number of them were little better than weeds with me. I do not wish to condemn them outright, for they may be attractive in other climates or locations, although some of them have been grown several times; so the names are not mentioned. Some time ago

I wrote, "I am convinced that some of these old garden species are not now worth growing for any horticultural ends." I have long wanted to publish the species traditionally grown for ornament or interest that should now be eliminated from cultivation and the names taken from the lists. Many new kinds have come into cultivation of much better quality and promise. We admire the sentiment attached to old kinds of things, but some garden plants have lost the charm of association and are no longer worth while. Progress is made by eliminating species that have lost their appeal as well as by adding new ones. And yet we must be cautious in making judgments, for plants that are indifferent with me may be excellent elsewhere. I have never had satisfaction with *Linaria maroccana*, for example, but recently saw it in rare perfection in spring in Southern California, grown in a great mass as a winter annual.

Considering all this interesting lot, assembled from the ends and corners of the earth, one wonders at the conservatism of the gardener; for the gardener desires the plants he or she has known before, perhaps those that were in the garden in the old home. So year after year we grow larkspurs and petunias, sweet peas, gladiolus, verbenas, calendulas, poppies, asters, candytufts, zinnias and marigolds, and a few others, content with the regularity, perhaps even the monotony. It is well to break away completely now and then, and take a season with new acquaintances.

It is not necessary for Americans to search for 307 kinds of plants in other lands. North America abounds in species good for planting of which no real or adequate use has yet been made. Laboriously we import plants from remote countries when our own woods and plains may bear kinds in every way as interesting and useful and much more in keeping with our situations; this is not because familiarity breeds contempt but because we are neither botanists nor horticulturists and therefore do not know our own resources. Schools disdain the study of the kinds of things, under the unnecessary prejudice that "systematic botany" is of no account, and instead may teach a kind of botany that has practically no significance to the pupil nor any superior training value. The generations are growing up in inexcusable ignorance of nature. Good appreciation and knowledge of horticulture must begin way back in the younger years and with the preliminary training.

THE 307 OF my garden-patch are gone. We knew them well, and enjoyed the acquaintance. We shall never see them assembled again; but if worthy of them we shall profit by the memory of them in the remaining days.

The Spirit of the Garden

I step from the house, and at once I am released. I am in a new realm. This realm has just been created, and created for me. I give myself over to the blue vault of the sky; or if it rain, to first-hand relationship with the elements,—for can I not touch the drops that fall from some mysterious height? I am conscious of a quick smell of the soil, something like the smell of the sea. I hear the call of a bird or a faint rush of wind, or catch a shadow that passes and is gone. There is a sudden sensation of green things tumbled over the ground. I feel that they are living, growing, aspiring, sensitive.

Then the details begin to grow up out of the area, every detail perfect in its way, every one individual, yet all harmonious. The late rain compacted the earth; but here are little grooves and cuts made by tiny rills that ran down the furrows and around the stems of the plants, coalescing and growing as they ran, digging gorges between mountainous clods, spreading into islanded lakelets, depositing deltas, and then plunging headlong toward some far-off sea,—a panorama that needs only to be magnified to make those systems of rivers and plains and mountains the names of which I sought so much in my old geography days.

Soft green things push up out of the earth, growing by some sweet alchemy that I cannot understand but that I can feel. Green leaves expand to the sun; buds burst into flowers; flowers change to fruits; the pods burst, and berries wither and fall; the seeds drop and are lost,—yet I know that nature the gardener will recover them in due season.

Strange plants that I did not want are growing here and there, and now I find that they are as good as the rest, for they spring from the same earth yet are unlike all others, they struggle for place and light, and they too will have their day and will die away and in some mysterious process will come again. Insects crawl here and there, coming from strange crevices and all of them intent. Earthworms heave their burrows. All these, too, pass on and die and will come again. A bird darts in and captures a flying insect; a dog trots across the farther end of the plot; a cat is hidden under the vines by the wall. A toad dozes under a bench: he will come out to-night.

It is all a drama, intense, complex, ever moving, always dying, always re-born. I see a thousand actors moving in and out, always going, always coming. I am part of the drama; I break the earth; I destroy this plant and that, as if I were the arbiter of life and death. I sow the seed. I see the tender things come up and I feel as if I had created something new and fine, that had not been seen on the earth before; and I have a new joy as deep and as intangible as the joy of religion.

Oak

Strength, solidity, durability are symbolized in the Oak. The tree is connected with the traditions of the race, and it is associated with literature. It is a tree of strong individuality, with bold, free growth and massive framework. Its longevity appeals to every person, even though he has no feeling for trees. It connects the present with the past. It spans the centuries.

This feeling that the Oak represents a long span of years is itself the reason why we should consider the tree with veneration and let it live its full time; and this is the particular lesson which the writer would impress. Spare the isolated Oak trees! Of whatever kind or species, a mature Oak is beyond price. To allow it to remain bespeaks culture and kindly feeling.

Many species of Oak are now available in nurseries. There are perhaps 25 species that can be relied on for planting in the northeastern states, and there are particular varieties adapted to almost every habitable part of North America. The planting of cheap, quick-growing willows and poplars is so common that one almost despairs of the time when such strong and expressive trees as Oaks shall be planted. There is little difficulty in the planting of Oaks if one secures nursery-grown stock. They grow more slowly than some other trees, but what they lack in rapidity of growth they make up in character and foliage color. If quick effects are wanted, some fast-growing tree may be planted with them, to be removed as the Oaks need the space. Some of the species grow nearly or quite as rapidly as hard maples, when young. Other species are mere bushes and make an excellent border-mass on the farther side of large grounds. Of such is the native Scrub-Oak (*Quercus ilicifolia*) of the eastern states. The native species are usually the best for any region, from the fact that they are adapted to climate and soil; and then, a feeling for common native plants is an indication of the highest appreciation and of the keenest response to the conditions in which one lives.

The Principles of Pruning

We are now prepared to enter on a discussion of some of the more underlying considerations governing the rational pruning of plants. It is a difficult field, for no two plants are alike, and many and various objects are to be attained. It is impossible to instruct any person in pruning by merely showing him how to do the work on a given plant; for the very next plant may present a new set of problems. If there are no generalizations or principles to be announced, then writing on the subject is well-nigh useless. The numbered statements or "principles" in this chapter are not intended to be dogmatic, for there are undoubtedly exceptions, or apparent exceptions, to all of them; but it is hoped that they separate some of the most important truths from the great mass of assertions and contradictions.

There are two great classes of ideas concerned in the pruning of plants—those which are associated directly with the welfare and behavior of the plant, and those which are associated with the mere form or size or convenience to which the plant shall attain. The former includes questions of pruning proper; the latter comprises questions of training, which depend primarily on the taste and abilities of the pruner. Shall I grow my trees to round heads or conical heads, high heads or low heads, one trunk or two trunks?

Whichever you like; it is largely a question of personal preference and opportunity.

Of all the operations connected with horticulture, pruning, shaping, and training bring the person into closest contact and sympathy with the plant. One directs and cares for the plant tenderly and thoughtfully, working out his ideals as he would in the training and guiding of a child. There are some persons, to be sure, who cannot feel this sympathetic contact with a plant: they are the ones who, if they prune at all, use an axe or machete or a corn-knife. If a person cannot love a plant after he has pruned it, then he has either done a poor job or is devoid of emotion. It is a pleasure to till the soil and to smell the fresh crumbly earth, but the earth does not grow; it is still a clod. The plant responds to every affectionate touch. Spraying, that modern warfare of horticulture, is not to be compared with pruning in producing a sense of fellowship with plants. In fact, spraying has the opposite effect with me. When I have sprayed a plant, I am conscious that I have besmeared it and have taken a mean advantage of a lot of innocent and defenseless bugs; and I want to quit the premises forthwith.

The reasons for pruning may be ranged under several general heads:

1. To modify the vigor of the plant.
2. To produce larger and better stems, leaves, flowers, or fruits.
3. To keep the plant within manageable shape and limits.
4. To change the habit of the plant from more or less wood-bearing or fruit-bearing (or flower-bearing).
5. To open a tree to light and air, for the betterment of the product.
6. To remove superfluous or injured parts.
7. To control the spread of disease, as in pear-blight.
8. To protect the plant against winds and snows, inasmuch as some ornamental trees and shrubs are particularly liable to injury if allowed to take their natural shape.

9. To expedite spraying and harvesting.
10. To facilitate tillage and to improve the convenience of the plantation.
11. To train the plant to some desired shape.

The Weather

That which is first worth knowing is that which is nearest at hand. The nearest at hand, in the natural surroundings, is the weather. Every day of our lives, on land or sea, whether we will or no, the air and the clouds and the sky environ us. So variable in this environment, from morning till evening and from evening till morning and from season to season, that we are always conscious of it. It is to the changes in this environment that we apply the folk-word "weather,"—weather, that is akin to wind.

No man is efficient who is at cross-purposes with the main currents of his life; no man is content and happy who is out of sympathy with the environment in which he is born to live: so the habit of grumbling at the weather is the most senseless and futile of all expenditures of human effort. Day by day we complain and fret at the weather, and when we are done with it we have—the weather. The same amount of energy put into wholesome work would have set civilization far in advance of its present state. Weather is not a human institution, and therefore it cannot be "bad." I have seen bad men, have read bad books, have made bad lectures, have lived two years about Boston,—but I have never seen bad weather!

"Bad weather" is mainly the fear of spoiling one's clothes. Fancy clothing is one of the greatest obstacles to a knowledge of nature:

in this regard, the farm boy has an immense advantage. It is a misfortune not to have gone barefoot in one's youth. A man cannot be a naturalist in patent-leather shoes. The perfecting of the manufacture of elaborate and fragile fabrics correlates well with our growing habit of living indoors. Our clothing is made chiefly for fair weather; when it becomes worn we use it for stormy weather, although it may be in no respect stormy weather clothing. I am always interested, when abroad with persons, in noting the various mental attitudes toward wind; and it is apparent that most of the displeasure from the wind arises from fear of disarranging the coiffure or from the difficulty of controling a garment.

If our clothes are not made for the weather, then we have failed to adapt ourselves to our conditions, and we are in worse state than the beasts of the field. Much of our clothing serves neither art nor utility. Nothing can be more prohibitive of an interest in nature than a millinery "hat," even though it be distinguished for its floriculture, landscape gardening, and natural history.

Our estimate of weather is perhaps the best criterion of our outlook on nature and the world. The first fault that I would correct in mankind is that of finding fault with the weather. We should put the child right toward the world in which he is to live. What would you think of the mariner who goes to sea only in fair weather? What have not the weather and the climate done for the steadiness and virility of the people of New England? And is this influence working as strongly today as in the times when we had learned less how to escape the weather? We must believe in all good physical comfort,—it contributes to the amount of work that we can accomplish; but we have forgotten that it is possible to bear an open storm with equanimity and comfort. The person who has never been caught in rain and enjoyed it has missed a privilege and a blessing.

Give us the rain and the hail and the snow, the mist, the crashing thunder, and the cold biting wind! Let us be men enough to face it, and poets enough to enjoy it. In "bad" weather is the time to go abroad in field and wood. You are fellow then with bird and stream and tree; and you are escaped from the crowd that is forever crying and clanging at your heels.

What Is a Weed?

From time to time attempts are made to find a suitable definition for "weed," as recently in *Science*. "Weed" is not a botanical term, and efforts to define it as a biological concept are futile. It is a folk-word and merely connotes an attitude of a person, who is trying to subdue the earth, toward any plant that bothers him or stands in his way. The word should not be disturbed from its general-language use. It may be explained but hardly defined. All one need do is to state what it signifies to farmers, gardeners and care-takers.

A weed is a plant that is not wanted by one who tills or dresses the earth. This simple and obvious statement I have used for many years. It occurs at least as early as in *Principles of Vegetable-Gardening*, p. 197 (1901). It occurs again in *Cyclopedia of American Horticulture*, 1902, p. 1972, as follows (and similarly in subsequent books):

> A weed is a plant that is not wanted. There are, therefore, no species of weeds, for a plant that is a weed in one place may not be in another. There are, of course, species that are habitual weeds; but in their wild state, where they do not intrude on cultivated areas, they can scarcely be called weeds. The common pigweed and the purslane are sometimes vegetables, in which case potato plants would be weeds if they grew among them.

Of course some plants are "weedier" than others, but this character does not make them weeds in the abstract. The statement takes a somewhat different form in Georgia's *Manual of Weeds* (one of the Rural Manual Series), 1914: "A weed is plant that is growing where it is desired that something else shall grow." The statement attributed to Emerson, that a weed is a plant out of place, has been much quoted; but it is a question whether a plant is ever out of place except when cultivated.

White Clover

A vagrant plant to my garden came
   And escaped the workman's hoe,
He knew it not by the leaf or name
   So he let it stay and grow.

It grew full well in the garden mould
   And covered a space yard-wide;
He watched the honey-white heads unfold
   And pointed them out with pride.

Many a weed in the garden lot
   Were fair as the clover blow
If only its name were all forgot
   And 'twere giv'n its chance to grow.

"Trifolium repens—*the White Clover.*"

III

FLOWERS

Flower loving is sentiment and emotion, kindled with imagination. It depends vastly more upon the person than it does the flower.

Blossoms

In January, 1912, a magazine called "Flowers" was set forth in New York. It was short-lived and I do not know when or why it came to its end. Recently the first page of the first issue was found among my papers, comprising "a foreword" by myself entitled "Blossoms." In part because a record may be made here of the magazine (and also in part because the essay has the merit of brevity), the article is reprinted here.

THERE ARE TWO parts to the common day,—the performance of the day, and the background of the day. Many of us are so submerged in the work we do and in the pride of life that the real day slips by unnoted and unknown. But there are some who part the hours now and then and let the background show through. There are others who keep the sentiments alive as an undertone and who hang all the hours of work on a golden cord, connecting everything and losing none; theirs is the full life; their backgrounds are never forgotten; and the backgrounds are the realities.

The joy of flowers is of the backgrounds. It lies deeper even than the colors, the fair fragrances, and the graces of shape. It is the joy of things growing because they must, of the essence of winds woven into a thousand forms, of a prophetic earth, and of wonderful delicateness in part and substance. The appeal is the deeper because we cannot analyze it, nor measure it by money, nor contain it in anything that we make with our hands. It is too fragile for analysis.

This fragile intimate brotherhood with the earth must always have been a powerful bond with men and an infinite resource to them, although we catch little of the feeling of it in the ancient literatures. Men must always have responded to the wild rose and to the tenderness of the grass. Certainly we know that men very early began to assemble blossoms about their homes, and to pass on the seeds from friend to friend.

Centuries ago great elaborate books were written about flowers, and the kinds even then were many and some of the forms were marvelous. Worship and praise have centered about flowers and garlands rather than about the fruits that we eat; this marks them to have been considered as of the higher things. All holy and great occasions need them if the occasions are complete. Not a soul but responds to blossoms, even though he knows it not. No soul passes a lily in blow, an apple orchard in the May, a clover field swept with red, or a good garden, but that some reflection of it enters his mind and lodges itself in some nexus of the brain. It would be difficult for any man to imagine a flowerless world; and if he conjured it in his dream he would find himself sitting in some oasis of greenery and bloom.

There is much speculation as to why flowers ever came into the world or what necessary utility they are to plants. But we are free to accept a fact; and flowers are facts. I think there must be something more than mere utility to the plant that brought blossoms into existence. But why ever they came, they are joyful things and they are parts in the journey in life. To know a flower well and to grow it well are more than botany and gardening. The songs of birds, the feel of the winds, the flow of the streams, the appeal of flowers, are so real that we are likely to forget them or to lose them; but the flowers excel them all in the ease and completeness with which we may adapt them to personal needs and incorporate them into a process of life.

The Symbolism of Flowers

Flowers are less valuable for their own sakes than for the emotions which they inspire. Their highest sphere is that of symbolism. So long and so intimately have we known this fact that we have all but forgotten it, and we often fail to distinguish the flower from the sentiment which it represents to us. We always think of flowers—unless we are analyzing them—as weak or bold, dull or gay, familiar or retiring. The attributes which appeal to us first are not the merely physical ones, as of redness, roundness or size. But as flowers cannot possess sentiments as a part of themselves, it follows that the emotions which they arouse are but a reflex of ourselves. One cannot see farther than his vision reaches, nor can he feel more than he has capacity for feeling. It is not strange, therefore, that the sweetest and most sympathetic minds find most to love in flowers. It must always be so; and yet, in some degree, flowers are symbols to all of us.

But can one learn to love flowers, then? Yes; and in the same way that he learns to love anything good and pure. If one could not outgrow himself there could be no such things as education and culture. But first of all, one must put himself in a sympathetic attitude with nature. Nature and man are one, not twain. Storms and floods, frosts and heat, shade and sun, are not man's enemies, but his allies. Consider yourself a part of nature, of fields and woods, and your ear at once becomes sympathetic and your heart learns new lessons of love. Then the little voices of nature speak to you, and you feel that you are in communion with the universe. All this does not come suddenly. Soon other men's thoughts add themselves to our own, and we find ourselves in touch with men and nature; and every flower suggests some thought or awakens hope.

Yet these symbolisms are spontaneous. We cannot construct them as we would carve a moulding or design a monogram. All history shows that they have arisen when least expected. The Scottish chiefs had no thought of the thistle until the Dane accidentally discovered it for them. So all our formal attempts to choose a national flower must perforce be failures. No one can conjure up a national sentiment nor bring about an occasion for the choice of a symbol. Ask an advocate of a national flower what the flower is to symbolize, and he will tell you that it is to symbolize the nation! But attributes, not states or objects, admit of symbols. Symbolize victory, charity, fortitude, but not the objects to which they belong; and it were a wretched symbol which were devised a century too late! No! If we choose a national flower, call it only a trademark for our escutcheon, not a symbol!

Extrinsic and Intrinsic
Views of Nature

"The purpose of this exercise is to tell children how to see the hidden beauties of flowers." Thus ran the announcement at the opening of the classroom period. Is it worthwhile to tell them any such thing? Why not teach them to be interested in plants? Why give them a half-truth when they might have the whole truth?

The "beauty" of a flower or a bird is only an incident: the plant or the bird is the important thing to know. Beauty is not an end. The person who starts out to see beauty in plants is often in the condition of mind that the dear old lady was who came into my conservatory and exclaimed, as she saw the geraniums, "Oh, they are as pretty as artificial flowers!"

But these people are not looking for beauty, after all; they look for mere satisfying form or color or oddity. They confound beauty with prettiness or with outward attractiveness. Real beauty is deeper than sensation. It inheres in fitness of means to end as well as in striking features. The child should see the object itself before he sees its parts or attributes. Teach first the whole bug, the whole bird, the whole plant, with something of the way in which it lives. The botanist may well devote his life to a cell, but the layman wants to know the trees and the woods.

I dislike to hear people say that they love flowers. They should love plants; then they have a deeper hold. Intellectual interest should go deeper than shape or color. Teachers or parents ask the child to see how "pretty" the object is; but in most cases the child wants to know how it lives and what it does.

It is instructive to note the increasing love for wild animals and plants as a country grows old and mature. This is particularly well illustrated in plants. In pioneer times there are too many plants. The effort is to get rid of them. The forest is razed and the road-sides are cleaned. The pioneer is satisfied with things in the gross. If he plants at all, he usually plants things exotic or strange to the neighborhood. The woman grows a geranium or fuchsia in a tin can, and now and then makes a flower-bed in the front yard; but the man is likely to think such things beneath him. If a man has flowers at all, he must have something that will fill the eye. Sun-flowers are satisfying.

But the second and third generations begin to plant forests and to allow the roadsides to grow wild at intervals. Persons come to be satisfied with their common surroundings and to derive less pleasure from objects merely because they are unlike their sur-roundings. Choice plants come into the yards here and there, and the men of the household begin to care for them. The birds and wild animals are cherished. (I know a man who in his pioneer days took no interest in crows except to get rid of them, but who later in life wept when a crow's nest in an apple tree was robbed.) Love of books increases. All this marks the growth of the intellectual and spiritual life.

America is a land of cut flowers. Nowhere does the cut-flower trade assume such commanding importance. Churches and homes are decorated with them. One sees the churches of the Old World decorated with plants in pots or tubs. The Englishman or the German loves to care for the plant from the time it sprouts until it dies: it is a companion. The American snips off its head and puts it in his buttonhole: it is an ornament. I have sometimes wondered whether the average flower-buyer knows that flowers grow on plants.

All of us have known persons who derive more satisfaction from a poor plant that never blooms than others do from a bunch of American Beauty roses at five dollars. There is individuality—I had almost said personality—in a growing, living plant, but there is little of it in a detached flower. And it does not matter so much if the plant is poor and weakly and scrawny. Do we not love poor and crippled and crooked people? A plant in the room on washday is worth more than a bunch of flowers on Sunday.

But the American taste is rapidly changing. Each year the florist's trade sees a proportionately greater demand for plants. The same change is seen in the parks and home grounds. Every summer more gross carpet-beds are relegated to those parts of the grounds that are devoted to curiosities, or they are omitted altogether, and in their stead are restful sward and attractive plant forms. Flowers are not to be despised, but they are accessories.

This habit of looking first at what we call the beauty of objects is closely associated with the old conceit that everything is made to please man: man is only demanding his own. It is true that everything is man's because he may use it or enjoy it, but not because it was designed and "made" for "him" in the beginning. This notion that all things were made for man's special pleasure is colossal self-assurance. It has none of the humility of the psalmist, who exclaimed, "What is man, that thou art mindful of him?"

"What were these things made for, then?" asked my friend. Just for themselves! Each thing lives for itself and its kind, and to live is worth the effort of living for man or bug. But there are more homely reasons for believing that things were not made for man alone. There was logic in the farmer's retort to the good man who told him that roses were made to make man happy. "No, they wa'n't," said the farmer, "or they wouldn't 'a' had prickers." A teacher asked me what snakes are "good for." Of course, there is but one answer: they are good to be snakes.

Being human, we interpret nature in human terms. Much of our interpretation of nature is only an interpretation of ourselves. Because a condition or a motive obtains in human affairs, we assume that it obtains everywhere. The only point of view is our own point of view. Of necessity, we assume a starting-point; therefrom we evolve an hypothesis which may be either truth or fallacy. Asa Gray combated Agassiz's hypothesis that species were originally created where we now find them and in approximately the same numbers by invoking Maupertuis's "principle of least action"—"that it is inconsistent with our idea of divine wisdom that the Creator should use more power than was necessary to accomplish a given end." The result may be secured with a less expenditure of energy than Agassiz's method would entail. But who knows that "our idea of the divine wisdom" is correct? It is only a human metaphor; but, being human, it may be useful.

Much of our thinking about nature is only the working out of propositions in logic, and logic is sometimes, I fear, but a clever substitute for truth. It is impossible to put ourselves in nature's place—if I may be allowed the phrase; that is, difficult to work from the standpoint of the organism that we are studying. If it were possible to get that point of view, it would be an end to much of our speculation; we should then deal with things as they are.

We hope that we are coming nearer to an intrinsic view of animals and plants; yet we are still so intent on discovering what ought to be, that we forget to accept what is.

The Flower-Growing Should Be Part of the Design

I do not mean to discourage the use of brilliant flowers and bright foliage and striking forms of vegetation; but these things are never primary considerations in a good domain. The structural elements of the place are designed first. The flanking and bordering masses are then planted. Finally the flowers and accessories are put in, as a house is painted after it is built. Flowers appear to best advantage when seen against a background of foliage, and they are then, also, an integral part of the picture. The flower-garden, as such, should be at the rear or side of a place, as all other personal appurtenances are; but flowers and bright leaves may be freely scattered along the borders and near the foliage masses.

It is a common saying that many people have no love or appreciation of flowers, but it is probably nearer to the truth to say that no person is wholly lacking in this respect. Even those persons who declare that they care nothing for flowers are generally deceived by their dislike of flower-beds and the conventional methods of flower-growing. I know many people who stoutly deny any liking for flowers, but who, nevertheless, are rejoiced with the blossoming of the orchards and the purpling of the clover fields. The fault may not lie so much with the persons themselves as with the methods of growing and displaying the flowers.

Defects in Flower-Growing

The greatest fault with our flower-growing is the stinginess of it. We grow our flowers as if they were the choicest rarities, to be coddled in a hotbed or under a bell-jar, and then to be exhibited as single specimens in some little pinched and ridiculous hole cut in the turf, or perched upon an ant-hill which some gardener has laboriously heaped upon a lawn. Nature, on the other hand, grows her flowers in the most luxurious abandon, and you can pick an armful without offense. She grows her flowers in earnest, as a man grows a crop of corn. One can revel in the color and the fragrance and be satisfied.

The next fault with our flower-growing is the flower-bed. Nature has no time to make flower-bed designs; she is busy growing flowers. And, then, if she were given to flower-beds, the whole effect would be lost, for she could no longer be luxurious and wanton, and if a flower were picked her whole scheme might be upset. Imagine a geranium-bed or a coleus-bed, with its wonderful "design," set out into a wood or in a free and open landscape! Even the birds would laugh at it!

What I want to say is that we should grow flowers freely when we make a flower-garden. We should have enough of them to make it worth the while. I sympathize with the man who likes sunflowers. There is enough of them to be worth looking at. They fill the eye. Now show this man ten square feet of pinks, or asters, or daisies, all growing free and easy, and he will tell you that he likes them. All this has a particular application to the farmer, who is often said to dislike flowers. He grows potatoes and buckwheat and weeds by the acre: two or three unhappy pinks or geraniums are not enough to make an impression.

Lawn Flower-Beds

The easiest way to spoil a good lawn is to put a flower-bed in it; and the most effective way in which to show off flowers to the least advantage is to plant them in a bed in the greensward. Flowers need a background. We do not hang our pictures on fence-posts. If flowers are to be grown on a lawn, let them be of the hardy kind, which can be naturalized in the sod and which grow freely in the tall unmown grass; or else perennials of such nature that they make attractive clumps by themselves. Lawns should be free and generous, but the more they are cut up and worried with trivial effects, the smaller and meaner they look.

But if we consider these lawn flower-beds wholly apart from their surroundings, we must admit that they are at best unsatisfactory. It generally amounts to this, that we have four months of sparse and downcast vegetation, one month of limp and frostbitten plants, and seven months of bare earth. I am not now opposing the carpet-beds which professional gardeners make in parks and other museums. I like museums, and some of the carpet-beds and set pieces are "fearfully and wonderfully made." I am directing my remarks to those humble home-made flower-beds which are so common in lawns of country and city homes alike. These beds are cut from the good fresh turf, often in the most fantastic designs, and are filled with such plants as the women of the place may be able to carry over in cellars or in the window. The plants themselves may look very well in pots, but when they are turned out of doors, they have a sorry time for a month adapting themselves to the sun and winds, and it is generally well on towards midsummer before they begin to cover the earth. During all these weeks they have demanded more time and labor than would have been needed to have cared for a plantation of much greater size and which would have given flowers every day from the time the birds began to nest in the spring until the last robin had flown in November.

Flower-Borders

We should acquire the habit of speaking of the flower-border. The border planting of which we have spoken sets bounds to the place, and makes it one's own. The person lives inside the place, not on it. Along the borders, against groups, often by the corners of the residence or in front of porches—these are places for flowers. Ten flowers against a background are more effective than a hundred in the open yard.

I have asked a professional artist, Mr. Mathews, to draw me the kind of a flower-bed that he likes. It is shown in the following figure. It is a border,—a strip of land two or three feet wide along a fence. This is the place where pigweeds usually grow. Here he has planted marigolds, gladiolus, goldenrod, wild asters, China asters, and—best of all—hollyhocks. Anyone would like that flower-garden. It has some of that local and indefinable charm which always attaches to an "old-fashioned garden" with its medley of form and color. Nearly every yard has some such strip of land along a rear walk or fence or against a building. It is the easiest thing to plant it,—ever so much easier than digging the characterless geranium bed into the center of an inoffensive lawn.

"An artist's flower-border."

The Old-Fashioned Garden

Speaking of the old-fashioned garden recalls one of William Falconer's excellent paragraphs ("Gardening," November 15, 1897, p. 75): "We tried it in Schenley Park this year. We needed a handy dumping ground, and hit on the head of a deep ravine between two woods; into it we dumped hundreds upon hundreds of wagon loads of rock and clay, filling it near to the top, then surfaced it with good soil. Here we planted some shrubs, and broadcast among them set out scarlet poppies, eschscholzias, dwarf nasturtiums, snapdragons, pansies, marigolds, and all manner of hardy herbaceous plants, having enough of each sort to make a mass of its kind and color, and the effect was fine. In the middle was a plantation of hundreds of clumps of Japan and German irises interplanted, thence succeeded by thousands of gladioli, and banded with montbretias, from which we had flowers till frost. The steep face of this hill was graded a little and a series of winding stone steps set into it, making the descent into the hollow quite easy; the stones were the rough uneven slabs secured in blasting the rocks when grading in other parts of the park, and both along outer edges of the steps and the sides of the upper walk a wide belt of moss pink was planted; and the banks all about were planted with shrubs, vines, wild roses, columbines, and other plants. More cameras and kodaks were leveled by visitors at this piece of gardening than at any other spot in the park, and still we had acres of painted summer beds."

Contents of the Flower-Beds

There is no prescribed rule as to what you should put into these flower-borders. Put in them the plants you like. Perhaps the greater part of them should be perennials that come up of themselves every spring, and which are hardy and reliable. Wild flowers are particularly effective. Everyone knows that many of the native herbs of woods and glades are more attractive than some of the most prized garden flowers. The greater part of these native flowers grow readily in cultivation, sometimes even in places which, in soil and exposure, are much unlike their native haunts. Many of them make thickened roots, and they may be safely transplanted at any time after the flowers have passed. To most persons the wild flowers are less known than many exotics which have smaller merit, and the extension of cultivation is constantly tending to annihilate them. Here, then, in the informal flower-border, is an opportunity to rescue them. Then one may sow in freely of easy-growing annuals, as marigolds, China asters, petunias and phloxes, and sweet peas.

One of the advantages of these borders is that they are always ready to receive more plants, unless they are full. That is, their symmetry is not marred if some plants are pulled out and others are put in. And if the weeds now and then get a start, very little harm is done. Such a border half full of weeds is handsomer than the average hole-in-the-lawn geranium bed. An ample border may receive wild plants every month in the year when the frost is out of the ground. Plants are dug in the woods or fields, whenever one is on an excursion, even if in July. The tops are cut off, the roots kept moist until they are placed in the border; most of these much-abused plants will grow. To be sure, one will secure some weeds; but then, the weeds are a part of the collection! Of course, some plants will resent this treatment, but the border may be a happy family, and be all the better and more personal because it is the result of moments of relaxation. Such a border has something new and interesting every month of the growing season; and even in the winter the tall clumps of grasses and aster-stems wave their plumes above the snow and are a source of delight to every frolicsome bevy of snowbirds.

I have spoken of a weedland to suggest how simple and easy a thing it is to make an attractive mass-plantation. One may make the most of a rock or bank, or other undesirable feature of the place. Dig up the ground and make it rich, and then set plants in it. You will not get it to suit you the first year, and perhaps not the second or the third; you can always pull out plants and put more in. I should not want a lawn-garden so perfect that I could not change it in some character each year; I should lose interest in it.

It must not be understood that I am speaking only for mixed borders. On the contrary, it is much better in most cases that each border or bed be dominated by the expression of one kind of flower or bush. In one place a person may desire a wild aster effect, or a petunia effect, or a larkspur effect, or a rhododendron effect; or it may be desirable to run heavily to strong foliage effects in one direction and to light flower effects in another. The mixed border is rather more a flower-garden idea than a landscape idea; when it shall be desirable to emphasize the one and when the other, cannot be set down in a book.

Annuals: The Best Kinds and How to Grow Them

A nnual plants are those that you must sow every year. From seed to seed is only a year or less. Annual plants probably comprise half the flowering plants of the world. They quickly take advantage of the moving seasons—grow, blossom, and die before they are caught by the blight of winter or of the parching dry season. They are shifty plants, now growing here, then absconding to other places. This very uncertainty and capriciousness makes them worth the while. The staid perennials I want for the main and permanent effects in my garden, but I could no more do without annuals than I could without the spices and the condiments at the table. They are flowers of a season: I like flowers of a season.

Of the kinds of annuals there is almost no end. This does not mean that all are equally good. For myself, I like to make the bold effects with a few of the old profuse and reliable kinds. I like whole masses and clouds of them. Then the other kinds I like to grow in smaller areas at one side, in a half-experimental way. There is no need of trying to grow equal quantities of all the kinds that you select. There is no emphasis and no modulation in such a scheme. There should be major and minor keys.

The minor keys may be of almost any kind of plant. Since these plants are semi-experimental, it does not matter if some of them fail outright. Why not begin the list at A and buy as many as you can afford and can accommodate this year, then continue the list next year? In five or ten years you will have grown the alphabet, and will have learned as much horticulture and botany as most persons learn in a college course. And some of these plants will become your permanent friends.

For the main and bold effects I want something that I can depend on. There I do not want to experiment. Never fill a conspicuous place with a kind of plant that you have never grown.

The kinds I like best are the ones easiest to grow. My personal equation, I suppose, determines this. Zinnia, petunia, marigold, four-o'clock, sunflower, phlox, scabiosa, sweet sultan, bachelor's-button, verbena, calendula, calliopsis, morning-glory, nasturtium, sweet pea—these are some of the kinds that are surest, and least attacked by bugs and fungi. I do not know where the investment of five cents will bring as great reward as in a packet of seeds of any of these plants.

Before one sets out to grow these or any other plants he must make for himself an ideal. Will he grow for a garden effect, or for specimen plants or specimen blooms? If for specimens, then each plant must have plenty of room and receive particular individual care. If for garden effect, then see to it that the entire space is solidly covered, and that you have a continuous maze of color. Usually the specimen plants would best be grown in a side garden, as vegetables are, where they can be tilled, trained, and severally cared for.

There is really a third ideal, and I hope that some of you may try it—to grow all the varieties of one species. You really do not know what the China aster or the balsam is until you have seen all the kinds of it. Suppose that you ask your seedsman to send you one packet of every variety of cockscomb that he has. Next year you may want to try stocks or annual poppies, or something else. All this will be a study in evolution.

There is still a fourth ideal—the growing for gathering or "picking." If you want many flowers for house decoration and to give away, then grow them at one side in regular rows as you would potatoes or sweet corn. Cultivate them by horse- or wheel-hoe. Harvest them in the same spirit that you would harvest string beans or tomatoes; that is what they are for. You do not have to consider the "looks" of your garden. You will not be afraid to pick them. The old stalks will remain, as the stumps of cabbages do. When you have harvested an armful your garden is not despoiled.

I like each plant in its season. China aster is a fall flower. In early summer I want pansies or candytufts and other early or quick bloomers. For the small amateur garden, greenhouses and hotbeds are unnecessary, and they are usually in the way. There are enough kinds of annuals that may be sown directly in the open ground, even in New York, to fill any garden. All those I have mentioned are such. In general, I should not try to secure unusually early effects in any kind of plant by starting it extra early. I should get early effects with kinds of plants that naturally are early. Let everything have its season. Do not try to telescope the months.

You can sow the seeds of most annuals even in May. I have sown China asters in the open ground in early June in New York State and have had excellent fall bloom. Things come up quickly and grow rapidly in May and June. They hurry. Don't expect to get spring bloom from annuals, but rather from perennials—the spring bulbs, soft bleeding-hearts, spicy pinks, bright-eyed polyanthuses and twenty more.

Make the soil rich and fine and soft and deep. There are some plants for which the soil can be made too rich, of course, but most persons do not err in this direction. For sweet peas there is this danger, for these are nitrogen gatherers, and the addition of nitrogenous manures makes them run too much to vine. The finer and more broken down the manure the better. Spade it in. Mix it thoroughly with the soil. If the soil is clay-like, see that fine manure is thoroughly mixed with the surface layer to prevent "baking."

Watering is an exacting labor, and yet half of it is usually unnecessary. The reasons why it is unnecessary are two: the soil is so shallowly prepared that the roots do not strike deep enough; we waste the moisture by allowing the soil to become hard, thereby setting up capillary connection with the atmosphere and letting the water escape. See how moist the soil is in spring. Mulch it so that the moisture will not evaporate. Mulch it with a garden rake, by keeping the soil loose and dry on top. This loose, dry soil is the mulch. There will be moisture underneath. Save water rather than add it. Then when you do have to water the plants, go at it as if you meant it. Do not dribble and piddle. Wet the soil clear through. Wet it at dusk or in cloudy weather. Before the hot sun strikes it, renew your mulch, or supply a mulch of fine litter. As many plants are spoiled by sprinkling as by drought. Bear in mind that watering is only a special practice; the general practice is so to fit and maintain the ground that the plants will not need watering.

The less your space the fewer the kinds you should plant. Have enough of each kind to be worth the while and the effort. It is as much trouble to raise one plant as a dozen.

It is usually best not to try to make formal "designs" with annuals. Such designs are special things, anyway, and should be used sparingly, and be made only by persons who are skilled in such work. A poor or unsuccessful design is the sorriest failure that a garden can have. Grow the plants for themselves—pinks because they are pinks, alyssum because it is alyssum, not because they may form a part of some impossible harp or angel.

This brings up a discussion of the proper place to put the annuals. Do not put them in the lawn: you want grass there, and grass and annuals do not thrive well together. Supposing that you grow the annuals for garden effect, there are two ways of disposing them—to grow in beds or in borders. Sometimes one method is better and sometimes the other. The border method is the more informal, and therefore the simpler and easier, and its pictorial effect is usually greater, but in some places there are no boundary lines that can be used for borders. Then beds may be used; but make the beds so large and fill them so full that they will not appear to be mere play-patches. Long beds are usually best. Four or five feet wide is about the limit of ease in working in them. The more elaborate the shape of the bed, the more time you will consume in keeping the geometry straight and the less on having fun with the plants. Long points that run off into the grass—as the points of a star—are particularly worrisome, for the grass-roots lock hands underneath and grab the food and moisture. A rectangular shape is best if you are intent only on growing flowers. Of course, if your heart is set on having a star on the lawn, you should have it; but you would better fill it with colored gravel.

It is surprising how many things one can grow in an old fence. The four-o'clocks shown illustrate this point.[1] Most persons owning this place would think that they had no room for flowers; yet there the four-o'clocks are, and they take up no room. Not all annuals will thrive under such conditions of partial neglect. The large-seeded, quick-germinating, rapid-growing kinds will do best. Sunflower, sweet pea, morning-glory, Japanese hop, zinnia, big marigold and amaranths are some of the kinds that may be expected to hold their own. If the effort is made to grow plants in such places, it is important to give them all the advantage possible early in the season, so that they will get well ahead of grass and weeds. Spade up the ground all you can. Add a little quick-acting fertilizer. It is best to start the plants in pots or small boxes, so that they will be in advance of the weeds when they are set out. First and last, I have grown practically every annual offered in the American trade. It is surprising how few of the uncommon or little-known sorts really have great merit for general purposes. There is nothing yet to take the place of the old-time groups, such as amaranths, zinnias, calendulas, daturas, balsams, annual pinks, candytufts, bachelor's-buttons, wallflowers, gilias, larkspurs, petunias, gaillardias, snapdragons, cockscombs, lobelias, coreopsis or calliopsis, California poppies, four-o'clocks, sweet sultans, phloxes, mignonettes, scabiosas, dwarf nasturtiums, marigolds, China asters, salpiglossis, nicotianas, pansies, portulacas, castor beans, poppies, sunflowers, verbenas, stocks, alyssums, and such good old running plants as scarlet runners, sweet peas, convolvuluses, ipomeas, nasturtiums, balloon vines and cobeas.[2] Of the annual vines of recent introduction, the Japanese hop has at once taken a prominent place for the covering of fences and arbors, although it has no floral beauty to recommend it.

For bold mass-displays of colour in the rear of the grounds or along the borders, some of the coarser species are desirable. My own favorites for such use are sunflower, castor bean, and striped Japanese corn for the back rows; zinnias for bright effects in the scarlets and lilacs; African marigolds for brilliant yellows; nicotianas for whites. Unfortunately, we have no robust-growing annuals with good blues. Some of the larkspurs are perhaps the nearest approach to it.

For lower-growing and less gross mass-displays the following are good: California poppies for oranges and yellows; sweet sultans for purples, whites and pale yellows; petunias for purples, violets and whites; larkspurs for blues and violets; bachelor's-buttons (or cornflowers) for blues; calliopsis and coreopsis and calendulas for yellows; gaillardias for red-yellows; China asters for many colors except yellow.

For still less robustness, good mass-displays can be made with the following: Alyssums and candytufts for whites; phloxes for whites and various pinks and reds; lobelias and browallias for blues; pinks for whites and various shades of pink; stocks for whites and reds and dull blues; wallflowers for brown-yellows; verbenas for many colours.

Some of the common annuals do not lend themselves well to mass-displays. They are of interest because of peculiar foliage, odd or unusual flowers, special uses, and the like. Of such are portulacas (for hot, sunny places), balsams, cockscombs, poppies (the blooming period is short), pansies, dwarf convolvuluses and dwarf nasturtiums, snapdragons, amaranths, four-o'clocks, mignonettes, alonsoas, schizanthus, nolanas, argemone, horned poppy, and many others.

I should never consider a garden of pleasant annual flowers to be complete that did not contain some of the "everlastings," or immortelles. These "paper flowers" are always interesting to children. I do not care for them for the making of "dry bouquets," but for their interest as a part of a garden. The colours are bright, the blooms hold long on the plant, and most of the kinds are very easy to grow. My favourite groups are the different kinds of xeranthemums and helichrysums. The gomphrenas, with clover-like heads (sometimes known as bachelor's-buttons), are good old favourites. Rhodanthes and ammobiums are also good.

Among the ornamental annual grasses, I have had most satisfaction with the brizas, coix or Job's tears, and some of the species of agrostis and eragrostis.

Some of the perennials and biennials can be treated as annuals if they are started very early indoors. A number of the very late-flowering annuals should also be started indoors for best success in the northern States, as, for example, the moonflowers and the tall-growing kinds of cosmos.

If flowers of any annual are wanted extra early, the seeds should be started indoors. It is not necessary to have a greenhouse for this purpose, although best results are to be expected with such a building. The seeds may be sown in boxes, and these boxes then placed in a sheltered position on the warm side of a building. At night they can be covered with boards or matting. In very cold "spells" the boxes should be brought inside. In this simple way seeds may often be started one to three weeks ahead of the time when they can be sown in the open garden. Moreover, the plants are likely to receive better care in these boxes, and therefore to grow more rapidly. Of course, if still earlier results are desired, the seeds should be sown in the kitchen, hotbed, cold frame, or in a greenhouse if accessible.

In starting plants ahead of the season, be careful not to use too deep boxes. The gardener's "flat" may be taken as a suggestion. Three inches of earth is sufficient, and in some cases (as when the plants are started late) half this depth is enough.

Of late years there has been a strong movement to introduce the hardy perennials into general cultivation. This is certainly to be encouraged everywhere, since it adds a feeling of permanency and purposefulness that is needed in American gardens. Yet I should be sorry if this movement were to obscure the importance of the annuals. We need this color and variety.

Campanula

There is a ferny dell I know
Where spiry stalks of harebell grow.
It is a little cool retreat
Of bosky scents and airs complete.
There is a maze of fragile stems
That hang their pods above the hems
Of mossy fountains crystal clear
'Mongst webby threads of gossamer
And filmy tints of green and blue
A-strung in beads of fragrant dew.
A tiny stroke the blue-bell rings
As on its slender cord it swings,
And if you listen long and well
You'll hear the music in the bell.

And often when I've toiled with men
Or passed my day with plans and pen
Or fled afar on starry seas,
I join the camp of moths and bees
And wander by the minty pools
To sedge and fern and campanules.
And then I lie on twig and grass
And watch the slimsy creatures pass,
And find the little folk that dwells

So deep inside the azure bells
I wonder how they come and go.
And as I listen long and low
I catch the cadence of a note
Astir within the petal throat,
I hear a tiny octave played
And slender music, crystal-rayed.

There are two worlds that I know full well—
The world of men and the petal bell.

IV

Fruits & Vegetables

I went from the market to the catalogues, and learned
that many things have been missed. I went from the
catalogues to the markets and found many things
as good as expected. In these circumstances I have
responsibility in the situation and should supplement
the markets with the products they cannot supply by
raising some of them myself.

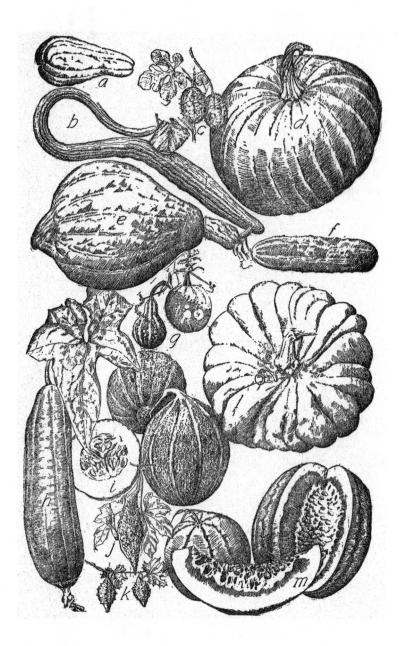

The Admiration of Good Materials

Not even yet am I done with this plain problem of the daily fare. The very fact that it is daily—thrice daily—and that it enters so much into the thought and effort of every one of us, makes it a subject of the deepest concern from every point of view. The aspect of the case that I am now to reassert is the effect of much of our food preparation in removing us from a knowledge of the good raw materials that come out of the abounding earth.

Let us stop to admire an apple. I see a committee of the old worthies in some fruit-show going slowly and discriminatingly among the plates of fruits, discussing the shapes and colors and sizes, catching the fragrance, debating the origins and the histories, and testing them with the utmost precaution and deliberation; and I follow to hear their judgment.

This kind of apple is very perfect in spherical form, deeply cut at the stem, well ridged at the shallow crater, beautifully splashed and streaked with carmine-red on a yellowish green under-color, finely flecked with dots, slightly russet on the shaded side, apparently a good keeper; its texture is fine-grained and uniform, flavor mildly subacid, the quality good to very good; if the tree is hardy and productive, this variety is to be recommended to the amateur for further trial! The next sample is somewhat elongated in form, rather below the average in color, the stem very long and well set and indicating a fruit that does not readily drop in windstorms, the texture exceedingly melting but the flavor slightly lacking in character and therefore rendering it of doubtful value for further test. Another sample lacks decidedly in quality, as judged by the specimens on the table, and the exhibitor is respectfully recommended to withdraw it from future exhibitions; another kind has a very pronounced aromatic odor, which will commend it to persons desiring to grow a choice collection of interesting fruits; still another is of good size, very firm and solid, of uniform red color, slightly oblate and therefore lending itself to easy packing, quality fair to good, and if the tree bears such uniform samples as those shown on the table it apparently gives promise of some usefulness as a market sort. My older friends, if they have something of the feeling of the pomologist, can construct the remainder of the picture.

In physical perfectness of form and texture and color, there is nothing in all the world that exceeds a well-grown fruit. Let it lie in the palm of your hand. Close your fingers slowly about it. Feel its firm or soft and modelled surface. Put it against your cheek, and inhale its fragrance. Trace its neutral under-colors, and follow its stripes and mark its dots. If an apple, trace the eye that lies in a moulded basin. Note its stem, how it stands firmly in its cavity, and let your imagination run back to the tree from which, when finally mature, it parted freely. This apple is not only the product of your labor, but it holds the essence of the year and it is in itself a thing of exquisite beauty. There is no other rondure and no other fragrance like this.

I am convinced that we need much to cultivate this appreciation of the physical perfectness of the fruits that we grow. We cannot afford to lose this note from our lives, for this may contribute a good part of our satisfaction of being in the world. The discriminating appreciation that one applies to a picture or a piece of sculpture may be equally applied to any fruit that grows on the commonest tree or bush in our field or to any animal that stands on a green pasture. It is no doubt a mark of a well-tempered mind that it can understand the significance of the forms in fruits and plants and animals and apply it in the work of the day.

I sometimes think that the rise of the culinary arts is banishing this fine old appreciation of fruits in their natural forms. There are so many ways of canning and preserving and evaporating and extracting the juices, so many disguises and so much fabrication, that the fruit is lost in the process. The tin-can and the bottle seem to have put an insuperable barrier between us and nature, and it is difficult for us to get back to a good munch of real apples under a tree or by the fireside. The difficulty is all the greater in our congested city life where orchards and trees are only a vacant memory or stories told to the young, and where the space in the larder is so small that apples must be purchased by the quart. The eating of good apples out of hand seems to be almost a lost art. Only the most indestructible kinds, along with leather-skinned oranges and withered bananas, seem to be purchasable in the market. The discriminating apple-eater in the Old World sends to a grower for samples of the kinds that he grows; and after the inquirer has tested them in the family, and discussed them, he orders his winter supply. The American leaves the matter to the cook and she orders plain apples; and she gets them.

I wonder whether in time the perfection of fabrication will not reach such a point that some fruits will be known to the great public only by the picture on the package or on the bottle. Every process that removes us one step farther from the earth is a distinct loss to the people, and yet we are rapidly coming into the habit of taking all things at second hand. My objection to the wine of the grape is not so much a question of abstinence as of the fact that I find no particular satisfaction in the shape and texture of a bottle.

If one has a sensitive appreciation of the beauty in form and color and modelling of the common fruits, he will find his interest gradually extending to other products. Some time ago I visited Hood River Valley in company with a rugged potato-grower from the Rocky Mountains. We were amazed at the wonderful scenery, and captivated by the beauty of the fruits. In one orchard the owner showed us with much satisfaction a brace of apples of perfect form and glowing colors. When the grower had properly expounded the marvels of Hood River apples, which he said were the finest in the world, my friend thrust his hand into his pocket and pulled out a potato, and said to the man: "Why is not that just as handsome as a Hood River apple?" And sure enough it was. For twenty-five years this grower had been raising and selecting the old Peachblow potato, until he had a form much more perfect than the old Peachblow ever was, with a uniform delicate pink skin, smooth surface, comely shape, and medium size, and with eyes very small and scarcely sunken; and my Hood River friend admitted that a potato as well as an apple may be handsome and satisfying to the hand and to the eye, and well worth carrying in one's pocket. But this was a high-bred potato, and not one of the common lot.

This episode of the potato allows me another opportunity to enforce my contention that we lose the fruit or the vegetable in the processes of cookery. The customary practice of "mashing" potatoes takes all the individuality out of the product, and the result is mostly so much starch. There is an important dietary side to this. Cut a thin slice across a potato and hold it to the light. Note the interior undifferentiated mass, and then the thick band of rind surrounding it. The potato flavor and a large part of the nutriment lie in this exterior. We slice this part away and fry, boil, or otherwise fuss up the remainder. When we mash it, we go still farther and break down the potato texture; and in the modern method we squeeze and strain it till we eliminate every part of the potato, leaving only a pasty mass, which, in my estimation, is not fit to eat. The potato should be cooked with the rind on, if it is a good potato, and if it is necessary to remove the outer skin the process should be performed after the cooking. The most toothsome part of the potato is in these outer portions, if the tuber is well grown and handled. We have so sophisticated the potato in the modern disguised cookery that we often practically ruin it as an article of food, and we have bred a race of people that sees nothing to admire in a good and well-grown potato tuber.

I now wish to take an excursion from the potato to the pumpkin. In all the range of vegetable products, I doubt whether there is a more perfect example of pleasing form, fine modelling, attractive texture and color, and more bracing odor, than in a well-grown and ripe field pumpkin. Place a pumpkin on your table; run your fingers down its smooth grooves; trace the furrows to the poles; take note of its form; absorb its rich color; get the tang of its fragrance. The roughness and ruggedness of its leaves, the sharp-angled stem strongly set, make a foil that a sculptor cannot improve. Then wonder how this marvellous thing was born out of your garden soil through the medium of one small strand of a succulent stem.

We all recognize the appeal of a bouquet of flowers, but we are unaware that we may have a bouquet of fruits. We have given little attention to arranging them, or any study of the kinds that consort well together, nor have we receptacles in which effectively to display them. Yet, apples and oranges and plums and grapes and nuts, and good melons and cucumbers and peppers and carrots and onions, may be arranged into the most artistic and satisfying combinations.

I would fall short of my obligation if I were to stop with the fruit of the tree and say nothing about the tree or the plant itself. In our haste for lawn trees of new kinds and from the uttermost parts, we forget that a fruit-tree is ornamental and that it provides acceptable shade. A full-grown apple-tree or pear-tree is one of the most individual and picturesque of trees. The foliage is good, the blossoms as handsome as those of fancy imported things, the fruits always interesting, and the tree is reliable. Nothing is more interesting than an orange tree, in the regions where it grows, with its shining and evergreen leaves and its continuing flowers and fruits. The practice of planting apples and pears and sweet cherries, and other fruit and nut trees, for shade and adornment is much to be commended in certain places.

But the point I wish specially to urge in this connection is the value of many kinds of fruit-trees in real landscape work. We think of these trees as single or separate specimens, but they may be used with good result in mass planting, when it is desired to produce a given effect in a large area or in one division of a property. I do not know that any one has worked out full plans for the combining of fruit-trees, nuts, and berry-bearing plants into good treatments, but it is much to be desired that this shall be done. Any of you can picture a sweep of countryside planted to these things that would be not only novel and striking, but at the same time conformable to the best traditions of artistic rendering.

I think it should be a fundamental purpose in our educational plans to acquaint the people with the common resources of the region, and particularly with those materials on which we subsist. If this is accepted, then we cannot deprive our parks, highways, and school grounds of the trees that bear the staple fruits. It is worth while to have an intellectual interest in a fruit-tree. I know a fruit-grower who secures many prizes for his apples and his pears; when he secures a blue ribbon, he ties it on the tree that bore the fruit.

The admiration of a good domestic animal is much to be desired. It develops a most responsible attitude in the man or the woman. I have observed a peculiar charm in the breeders of these wonderful animals, a certain poise and masterfulness and breadth of sympathy. To admire a good horse and to know just why he admires him is a great resource to any man, as also to feel the responsibility for the care and health of any flock or herd. Fowls, pigs, sheep on their pastures, cows, mules, all perfect of their kind, all sensitive, all of them marvellous in their forms and powers,— verily these are good to know.

If the raw materials grow out of the holy earth, then a man should have pride in producing them, and also in handling them. As a man thinketh of his materials, so doth he profit in the use of them. He builds them into himself. There is a wide-spread feeling that in some way these materials reflect themselves in a man's bearing. One type of man grows out of the handling of rocks, another out of the handling of fishes, another out of the growing of the products from the good earth. All irreverence in the handling of these materials that come out of the earth's bounty, and all waste and poor workmanship, make for a low spiritual expression.

The farmer specially should be proud of his materials, he is so close to the sources and so hard against the backgrounds. Moreover, he cannot conceal his materials. He cannot lock up his farm or disguise his crops. He lives on his farm, and visibly with his products. The architect does not live in the houses and temples he builds. The engineer does not live on his bridge. The miner does not live in his mine. Even the sailor has his home away from his ship. But the farmer cannot separate himself from his works. Every bushel of buckwheat and every barrel of apples and every bale of cotton bears his name; the beef that he takes to market, the sheep that he herds on his pastures, the horse that he drives,—these are his products and they carry his name. He should have the same pride in these—his productions—as another who builds a machine, or another who writes a book about them. The admiration of a field of hay, of a cow producing milk, of a shapely and fragrant head of cabbage, is a great force for good.

It would mean much if we could celebrate the raw materials and the products. Particularly is it good to celebrate the yearly bounty. The Puritans recognized their immediate dependence on the products of the ground, and their celebration was connected with religion. I should be sorry if our celebrations were to be wholly secular.

We have been much given to the display of fabricated materials,— of the products of looms, lathes, foundries, and many factories of skill. We also exhibit the agricultural produce, but largely in a crass and rude way to display bulk and to win prizes. We now begin to arrange our exhibitions for color effect, comparison, and educational influence. But we do not justly understand the natural products when we confine them to formal exhibitions. They must be incorporated into many celebrations, expressing therein the earth's bounty and our appreciation of it. The usual and common products, domesticated and wild, should be gathered in these occasions, and not for competition or for prize awards or even for display, but for their intrinsic qualities. An apple day or an apple sabbath would teach the people to express their gratitude for apples. The moral obligation to grow good apples, to handle them honestly, to treat the soil and the trees fairly and reverently, could be developed as a living practical philosophy into the working-days of an apple-growing people. The technical knowledge we now possess requires the moral support of a stimulated public appreciation to make it a thoroughly effective force.

Many of the products and crops lend themselves well to this kind of admiration, and all of them should awaken gratitude and reverence. Sermons and teaching may issue from them. Nor is it necessary that this gratitude be expressed only in collected materials, or that all preaching and all teaching shall be indoors. The best understanding of our relations to the earth will be possible when we learn how to apply our devotions in the open places.

The Affection for the Work

This inventory contains all the leading vegetable-garden plants of the world, and a good number of those of minor importance.[1] It suggests the variety and wealth of the field in plant materials. It would run into many hundreds more if a complete list were attempted. In 1889, Sturtevant (Agric. Sci. iii: 174–8) classified 1,070 species of cultivated food plants, and added that his notes include 4,233 species of edible plants in 1,353 genera and 170 families.[2] These plants comprise all classes,—grains, fruits, vegetables and others. Undoubtedly these numbers could now be much increased.

In the foregoing lists are 247 entries, of which 114 are leaf vegetables, 59 root vegetables, and 74 fruit vegetables. It displays a fascinating field for labor and study. Here are seeds of unimagined forms, oddities in germination, growths to fix the attention, flowers and fruits representing the vast range of the vegetable kingdom, products in which one may take a personal pride. The number of domesticated forms is sumless, and yet the opportunity for plant-breeding is without end. Who knows the fruits of even the common vegetables? Who can describe accurately even one of the plants, as the botanist would describe it if he had his material properly preserved before him? Where are the herbaria and the museums in which the common things, to say nothing of the uncommon ones, are adequately collected? Plant-growing is so commercialized that we are tempted to give most of our attention to the mechanical and business aspects of the subject, losing our skill as plantsmen. But whatever the development of any one of these industries, we must remember that the starting-point is the seed, and that the horticulturist must ever renew his effort to get back to the plant. This effort is not to be conceived as an impersonal task yielding results for commerce and science, but as an ardent affection.

This affection runs not only to the growing of the plants and to the joy of gardening, but also to the appreciation of the good quality that one gets directly from fresh vegetables of merit. It is good to know the plants on which these products grow. As millions of people do not have gardens, so are they unaware of the low quality of much of the commercial produce as compared with things well grown in due season. Most persons, depending on the market, do not know what a superlative watermelon is like. Even such apparently indestructible things as cucumbers have a crispness and delicacy when taken directly from the vine at proper maturity that are lost to the store-window supply. Every vegetable naturally loses something of itself in the process from field to consumer. When to this is added the depreciation by storage, careless exposure and rough handling, one cannot expect to receive the full odor and the characteristic delicacies that belong to the product in nature. We must also remember the long distances over which much of the produce must be transported, and the necessity to pick the produce before it is really fit, to meet the popular desire to have vegetables out of season and when we ought not to want them. There is a time and place for everything, vegetables with the rest. Modern methods of marketing, storing and handling have facilitated transactions, and they have also done very much to safeguard the produce itself and to deliver it to the customer in good condition; but the vegetable well chosen and well grown and fresh from the garden is nevertheless the proper standard of excellence. It is a surpassing satisfaction when the householder may go to her own garden rather than to the store for her lettuce, onions, tomatoes, beets, peas, cabbage, melons, and other things good to see and to eat, and to have them in generous supply.

The Growing of the Vegetable Plants

"Cultivating the backache."

A vegetable garden is admittedly a part of any home place that has a good rear area. A purchased vegetable is never the same as one taken from a man's own soil and representing his own effort and solicitude.

It is essential to any satisfaction in vegetable-growing that the soil be rich and thoroughly subdued and fined. The plantation should also be so arranged that the tilling can be done with wheel tools, and, where the space will allow it, with horse tools.

The old-time garden bed (seen above) consumes time and labor, wastes moisture, and is more trouble and expense than it is worth. The rows of vegetables should be as long and continuous as possible, to allow of tillage with wheel tools. If it is not desired to grow a full row of any one vegetable, the line may be made up of several species, one following the other, care being taken to place together such kinds as have similar requirements; one row, for example, might contain all the parsnips, carrots, salsify. One or two long rows containing a dozen kinds of vegetables are usually preferable to a dozen short rows, each with one kind of vegetable.

It is well to place the permanent vegetables, as rhubarb and asparagus, at one side, where they will not interfere with the plowing or tilling. The annual vegetables should be grown on different parts of the area in succeeding years, thus practicing something like a rotation of crops. If radish or cabbage maggots or club-root become thoroughly established in the plantation, omit for a year or more the vegetables on which they live.

A suggestive arrangement for a kitchen-garden is given on page 139. On page 140 is a plan of a fenced garden, in which gates are provided at the ends to allow the turning of a horse and cultivator (Webb Donnell, in *American Gardening*). Page 141 shows a garden with continuous rows, but with two breaks running across the area, dividing the plantation into blocks. The area is surrounded with a windbreak, and the frames and permanent plants are at one side.

It is by no means necessary that the vegetable-garden contain only kitchen-garden products. Flowers may be dropped in here and there wherever a vacant corner occurs or a plant dies. Such informal and mixed gardens usually have a personal character that adds greatly to their interest, and, therefore, to their value. One is generally impressed with this informal character of the home-gardens in many European countries, a type of planting that arises

| 6 ft. | 6 ft. | 4 ft. | 4 ft. | 3 ft. | 3 ft. | 2½ ft | 2½ ft. | 2½ ft. | 4 ft. | 4 ft. | 4 ft. | 4 ft. | 6 ft. | 8 ft. | 8 ft. |

Asparagus.

Rhubarb.

Artichoke.

Parsnip.

Salsify.

Cucumbers, followed by Fall Spinach.

Peas

Early Potatoes or Peas, followed by Celery.

Early Cabbage and Cauliflower.

Beets. — Turnips.

Lettuce, early and late. — Winter Radish. — Endive. — Parsley.

Onions, with early Radish sown in row.

Bush Beans.

Late Cabbage.

Early Corn and Summer Squash.

Late Corn.

Tomatoes and Pole Beans.

Musk and Watermelon.

Winter Squash.

"Tracy's plan for a kitchen-garden."

"A garden fence arranged to allow horse work."

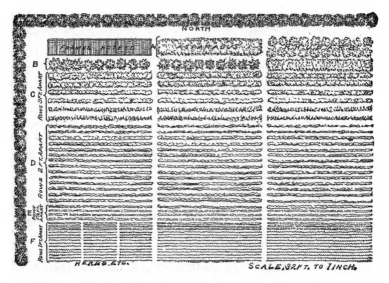

"A family kitchen-garden."

from the necessity of making the most of every inch of land. It was the writer's pleasure to look over the fence of a Bavarian peasant's garden and to see, on a space about 40 feet by 100 feet in area, a delightful medley of onions, pole beans, peonies, celery, balsams, gooseberries, coleus, cabbages, sunflowers, beets, poppies, cucumbers, morning-glories, kohl-rabi, verbenas, bush beans, pinks, stocks, currants, wormwood, parsley, carrots, kale, perennial phlox, nasturtiums, feverfew, lettuce, lilies!

The Fruit-Garden

I bring tokens from the middle of the past century.

The subject of this treatise is one in which almost all classes of the community are more or less practically engaged and interested. Agriculture is pursued by one class, and commerce by another; the mechanic arts, fine arts, and learned professions by others; but fruit culture, to a greater or less extent, by all.

It is the desire of every man, whatever may be his pursuit or condition in life, whether he live in town or country, to enjoy fine fruits, to provide them for his family, and, if possible, to cultivate the trees in his own garden with his own hands. The agriculturist, whatever be the extent or condition of his grounds, considers an orchard, at least, indispensable. The merchant or professional man who has, by half a lifetime of drudgery in town, secured a fortune or a competency that enables him to retire to a country or suburban villa, looks forward to his fruit garden as one of the chief sources of those rural comforts and pleasures he so long and so earnestly labored and hoped for. The artizan who has laid up enough from his earnings to purchase a homestead, considers the planting of his fruit trees as one of the first and most important steps towards improvement. He anticipates the pleasure of tending them in his spare hours, of watching their growth and progress to maturity, and of gathering their ripe and delicious fruits, and placing them before his family and friends as the valued products of his own garden, and of his own skill and labor. Fortunately, in the United States, land is so easily obtained as to be within the reach of every industrious man; and the climate and soil being so favorable to the production of fruit, Americans, if they be not already, must become truly "a nation of fruit growers."

Fruit culture, therefore, whether considered as a branch of profitable industry, or as exercising a most beneficial influence upon health, habits, and tastes of the people, becomes a great national interest, and whatever may assist in making it better adapted to the various wants, tastes, and circumstances of the community, cannot fail to subserve the public good.

The quotation above is from the introduction to the original edition of Patrick Barry's "Fruit Garden." This was a notable book of its day. It was published in New York by Charles Scribner in 1851, again in 1857 in Auburn and Rochester, New York by Alden & Beardsley, in 1860 again in New York City by C. M. Saxton, Barker & Co., and in 1863 in Rochester by the same publishers, in 1872 in New York City by the Orange Judd Company and again by the same in 1883, and for aught we know at other times; all this attests to its value and its popularity. With all the burning enthusiasms of the present day in the commercial conquests of the fruit men, we can hardly match the fervor of this artistic piece of writing; and in comparison with it much of the contemporaneous writing, compounded on the typewriting machine, lacks sadly in color and personality.

Evidently there was a spontaneity in that epoch that is impossible of attainment in these times of institutionalism and standardization. Probably it is not alone the memories of youth that give those days their flavor to me. There was something very real in the animation of that period, when the frontiers were expanding, when hope was in every migration, and when every new thing was tried and persons were not obliged to wait for official tests. To be sure, I do not go back to the first edition of "The Fruit Garden," but it was my great happiness to have known Patrick Barry, Charles Gibb, John J. Thomas, Marshall P. Wilder, C. M. Hovey, and others whose lives did not reach the present century. There was a certain masterfulness in those men, born of long experience and varied interests. Most of our first great horticulturists were fruit men although not to the exclusion of other attachments. Their names have come to us as symbols of authority, and yet their reputations were made on small areas or at least in activities that would now seem to be circumscribed. They were amateurs in the best sense. Consider Wilder's interest in his importations of the best pears and camelias, yet a business man in Boston. Turn over the pages of Hovey's Magazine of Horticulture. Note Gibb's introductions of the Russian fruits. Remember that Thomas wrote on many subjects. And Barry's "Fruit Garden" is directly a book for the amateur; probably not yet has a work on commercial fruit-growing exerted such an influence.

Consider also the Downings, who stand in the forefront of American horticulture. I remember my father writing to Charles Downing from what was then considered to be the western wilderness asking for cions of the Surprise apple that he might add it to the many varieties growing on his place and offering to pay for them; and Downing sent the cions and wrote in his own hand (there were no typewriters then) that there was no charge to those who loved fruits. I set those cions; and for many a year made pilgrimage to the tree and opened the green fruits to be surprised again and again at the pink flesh, "stained with red" as the original "Fruits and Fruit Trees of America" has it; and indeed it was "of little or no value," as the account says, but father and I knew that it paid its keep (there was no cost-accounting then). I have not seen this variety in any number of years; if any man still grows it, will he not send me a fruit when it is ripe?—and I shall not even offer to pay for it.

An instructive history may one day be prepared from the lives of the masters in American pomology. These lives had quality and breadth of view, but the main contribution was the fact that they associated fruits with the home life. I cannot forget Charles Downing's fruit-house which I early photographed and of which an illustration was made for the first Cyclopedia of Horticulture and repeated in the second. Here was a place in which numbers of varieties of fruits could be kept, always accessible for "testing" on

"Charles Downing's fruit-house."

occasion. The early exhibitions of fruits turned mostly on entries of "plates" rather than on "packs," and the samples were painstakingly judged for excellence and quality; and the judgments were opinions of chosen men, not footings of scorecards. One wonders whether there ever will be a recrudescence of this kind of pomology?

PATRICK BARRY WROTE of fruit-growing for the home garden. He would expound "the art of planting fifty trees on a quarter of an acre of ground, and bringing them into a fruitful state in four or five years at most." He begins the book with a chapter on "names, descriptions, and offices of the different parts of fruit trees," to the end that the grower might have close and sympathetic relations with the tree itself. Now, when we write books about crops, we are likely to begin with census figures.

The "new planting spirit" was abroad when Barry wrote; "until within a few years nothing was said or known among the great body of cultivators, or even nursery-men, of dwarfing trees." Yet he declares that "nothing so distinguishes the taste of modern planting as the partiality for dwarf trees, and the desire to obtain information in regard to their propagation and treatment," and he set out to supply information "in regard to the management of trees in the more refined and artificial forms."

Except in pears, the demand for dwarf fruit-trees was apparently not of long duration. The dwarf pear tree was a commonplace in my boyhood and many excellent plantations were in their prime when I entered my teaching work; today dwarf pears are offered little by nurserymen, nor, in fact, few pears of any kind outside a very few commercial sorts, and they have largely gone out in the East. Hedrick remarks in the notable "Pears of New York," published in 1921 as a report from the New York Agricultural Experiment

Station, that "a dwarf orchard or even a dwarf tree is now seldom seen." A worthy effort to revive and extend an interest in dwarf fruit-trees was made in 1906 by Waugh in a book on the subject.

But the trend of gardening enterprise runs to roses, irises, peonies, rock-gardens and many other colorful subjects and the fruit-garden of Barry's day is yet essentially unrealized as a recognized effort. This is a vast pity, for not only is the fruit itself of prime interest and value but the care required in the rearing of it brings one into the most intimate relationship with the plant and should afford a singular mental training. Nothing more clearly shows the skill and devotion of the horticulturist than the ability to grow good fruit in a small space, even though the subject is not essentially difficult; and the fact that such fruit-raising is not the vogue should be specially suggestive to one who desires to find a personal expression.

There was probably greater interest in the planting of fruits when Barry wrote and shortly afterward than ever before in North America. He attributes the beginning of this impulse largely to Downing's "Fruits and Fruit Trees of America," 1845, "the first treatise of the kind that really obtained a wide and general circulation." This book made its appearance, he states "at a favorable moment, just as the planting spirit referred to was beginning to manifest itself, and when, more than at any previous period, such a work was needed. Mr. Downing enjoyed great advantages over any previous American writer. During the ten years that had elapsed since the publication of Kenrick's and Prince's treatises, a great fund of materials had been accumulating. Messrs. Manning, Kenrick, Prince, Wilder, and many others, had been industriously collecting fruits both at home and abroad. The Massachusetts Horticultural Society was actively engaged in its labors. The London Horticultural Society had made great advancement in its examination and trial of fruits, and had corrected a multitude of long-standing errors in nomenclature. Mr. Downing's work had the benefit of all this; and possessing the instructive feature of outline figures of fruits, and being written in a very agreeable and attractive style, it possessed the elements of popularity and usefulness in an eminent degree."

Barry gives just praise also to Thomas' "American Fruit Culturist," which first appeared at Auburn, New York, near which Mr. Thomas lived, in 1849. Cole's "American Fruit-Book," 1849, is also commended. "Besides these, periodicals devoted more or less to the subject, have increased in number and greatly extended their circulation, so that information is now accessible to all who desire it." Hovey's "Magazine of Horticulture" and Downing's "Horticulturist" were then in full life. Meehan's "Gardeners' Monthly" began in 1859. Purdy's "Fruit Recorder," which undoubtedly had great influence on fruit-culture, began in 1869 and continued until 1886. But now it cannot be said that the agriculturist, "whatever be the extent or condition of his grounds," considers an orchard indispensable; nor does the merchant, professional man and artizan commonly contemplate the planting of fruit-trees. Home-production of all kinds has mostly gone out. Fruit-growing, like other enterprises, has become specialized. The fruit market is supplied regularly in standardized sizes and containers. It follows that the variety has been reduced, and if one wants fruit of a particular and superior kind one must expect to grow it. We have felt that a recrudescence of home fruit-culture would stimulate the market for fruit by increasing the knowledge and desire of it and incite a wholesome emulation and thereby be to the advantage of both the home grower and the commercial grower.

MANY YEARS AGO I planted rather heavily of dwarf apples and they did well as long as they received particular care. Two hundred trees were imported from France. Many others were propagated to choice kinds on imported Paradise-apple roots, and these were most satisfactory because the varieties were better suited to my needs. For years they bore heavily and the fruit was superior, as my friends and the market testified. Most of the trees were set between standard apples with the idea of removing them when the larger trees needed the room. They are apparently as permanent, however, as standard or "free" trees if properly cared for. They are not to be planted in this country for commercial apple-orcharding, but they are readily grown and are capable of giving much satisfaction to the amateur. One should not attempt to grow common commercial apples on the dwarf stocks. Only choice dessert varieties should find place in the dwarf orchard, and the total yield is less important than the high quality of the product. The only published result of my work with the dwarf apples is Waugh's vivid (and truthful) comment that I had had a lot of experience but did not know what it was.

One develops a sense of permanency when one plants fruit-trees, particularly of the long-enduring kinds as sweet cherries, pears, and apples; and with the other tree-fruits and the dwarfs one feels that the work is not for a season or a year but for a long term, perhaps even for a lifetime. In 1760 John Bartram planted in his garden on the Schuylkill a pear tree sent from England by Lady Petre; more than one hundred and twenty-five years later I took cions from the tree and set them in a young pear of my planting, and the tree is still fruitful. Plantations of some of the bush-fruits last any number of years if well cared for. A fruit-growing effort is peculiarly adapted to the settled home, where one expects to abide. The years come and go, and the trees hold their place and grow better and more characteristic with time; the element of maturity is in them. The spring will see the bloom, and another and another spring will see it again on the same tree, the tree grown older, more fruitful and more worth the while. It is the pride of the grower to make sure that it is more worth while. The autumn year after year will fulfill the promise of the year, and the product will fill the baskets.

OF EVERY TREE, whether pear tree or elm or sassafras, the seasons make a harbinger. I knew two oak trees intimately for some years, seeing them practically every day. I knew their blooming in relation to each other and the shedding of their leaves. Their autumn colors were peculiar shades and every year of the same quality each of itself. I wrote of them one autumn long ago: "the oak on one side my doorway is maroon-red and that on the other side is veiny-yellow, and they have been the same in all the Octobers in which I have loved them." And now my elms and maples are the same to me. Some of them leaf out late in spring, some early, and they shed their leaves in something like a succession. We do not know why; perhaps that is one reason why we like them so well. The fruit-tree becomes more distinctly an entity because one knows it so much better by closer care of it. Its vicissitudes and its welfare are more particularly ours. The love of a tree, especially if identified with a home, is one of the sacred associations.

Some twenty-five years ago I wrote my appreciation of a certain elm tree that tragedy had overtaken and printed it in Country Life in America.[3] The manuscript of it is now before me, and the theme of it so well suits the present purpose that it is placed here to complete the sentiment; for a fruit-tree is kin to every other tree.

In a long row of noble roadside maples was a single elm tree. It was planted where a maple had died. With large maples on either side and an orchard at the back, it had to struggle for light and moisture and food. Younger and smaller than the other trees, it was the special attraction of roaming cattle and the particular mark of mischievous children. The soil was too dry to suit an elm tree. Yet, in spite of all these disabilities, the tree grew and flourished.

Every passing storm seemed to wreak its vengeance on this elm tree. One late winter morning we awoke to find the world transformed by ice on every tree and bush. In wonder and amazement we looked abroad. But in front of us the elm tree lay a shapeless mass, broken and splintered by the weight of ice. Already the tree had been endeared to us by its many hardships, in which all the family had sympathized. The tree must not perish now.

With ropes and pulleys the great limbs, some of them now several inches in diameter, were drawn back to their places; for every one of them still clung to the parent stock by a strip of bark and wood at its base. Iron bolts were made from half-inch rods, long enough to reach through branch and stock just above the split. With long auger, half-inch holes were bored through the tree, the bolts driven in tight and then drawn up by means of a nut and thread. A large head and washer, and another washer under the nut at the other end, prevented the ends of the bolt from drawing into the wood. So firmly was the branch drawn to the trunk that no gaping crack was left, and the crease was hermetically sealed with melted wax. Then higher up, between branches two or three feet apart, other rods were run to hold all the members in place. We knew that if the bolts fitted tight in their holes, no harm would come to the tree; but that if bands were placed about the branches they would soon crease and girdle the parts and work much injury. When the storm had passed, the tree stood in its customary mood; and all the following summer it grew as if with renewed determination.

Nor was this the last of its misfortunes. Again an ice storm severed the great head until its branches lay prone on the ground. Now it was necessary to reduce the weight of the fallen branches, so long and large, by removing several feet of their ends. We knew that this cutting back would also tend to concentrate the new growth near the base and therefore to lessen the chances of another wrecking until the splits were well healed. Then again the parts were raised and bolted, and again the tree took up its thread of life undaunted. The old man who cared for this tree had fought his way through many obstacles; but he did not know why it was so dear to him.

The old tree is two feet in diameter now. Its iron bolts are all overgrown and no visible trace of them remains. It is no longer protected by cattle-guards. It has overtopped the maples. Birds build their nests in it. Now and then I wander along that old roadside. I admire the avenue of great maples and hear the praises of them from fellow travelers. But when I come to the old elm tree, almost unconsciously I throw an arm around its sturdy bole and bury my fingers in the furrows of its bark.

BUT IT IS the fruit-tree that is specially celebrated on this occasion. The fruit-tree differs in the fact that it has been so propagated that it bears a particular variety of fruit, and it should be more carefully pruned than are the trees grown for ornament or shade. The seed from which the tree made its start reproduces the species only, *Prunus avium* if a sweet cherry, *Pyrus communis* if a pear, *Citrus sinensis* if a sweet orange. But the particular variety, whether Windsor or Napoleon, Bartlett or Seckel, Navel or Valencia, lies in the wood rather than in the seed, and a cion or cutting of it is inserted in the seedling stock to become the bearing part of the future tree. Sometimes the plant grows straight from a cutting, as the grape, and directly produces the given variety. I am not making these statements for the unnecessary purpose of giving information but to illustrate how close is the relation between a fruit-tree and a man.

Yet it is in the pruning that this relation is best expressed. Day after day I have been alone with the trees, with a knife, shears and a light saw, and have convinced them into the ways that they should go. The trunk is to be so tall; the main or scaffolding limbs just so many; the height and shape of the future head are anticipated; the branches that will interfere are removed. Ten years from now the tree will express my handiwork and my judgment. Let us trust the result will be good.

The reader has been so patient with me for having introduced an old article that I venture to repeat the indiscretion. Before me is a manuscript on "the fruit-tree" that surely was written years ago and which probably was published but I do not know where, perhaps for children somewhere.[4] We may use at least part of it. The essay must have been written when my pen was nimbler than now, and if the reader does not like it he has the alternative of passing it by; and that will close my plea for the fruit-garden.

By the woodshed or the pump, or against the barn or the garden fence, the apple tree or pear tree connects the residence with the world of life and space that stretches out to woods and farms. We transfer our affections to it, as a half-way place between ourselves and our surroundings. It is the warder of the fields and the monitor of the home. It is an outpost of the birds. It feels the first ray of morning sunshine. It proclaims every wind. It drips copiously in the rain. Its leaves play on the grass when the year goes down into the long night of winter. And in the spring its brightening twigs and swelling buds reveal the first pulse in the reviving earth. Every day of the year is in its fabric, and every essence of wind and sun and snapping frost is in its blossom and its fruit.

I often wonder what must have been the loss of the child that had no fruit-tree to shelter it. There are no memories like the days under an old apple tree. Every bird of the field comes to it sooner or later. Perhaps a humming-bird once built on the top of a limb, and the marks of the old nest are still there. Strange insects are in its knots and wrinkles. The shades are very deep and cool under it. The sweet smells of spring are sweetest there. And the mystery of the fruit that comes out of a blossom is beyond all reckoning, the magic growing week by week until the green young balls show themselves gladly among the leaves. And who has not watched for the first red that comes on the side that hangs to the sun, and waited for the first fruit that was soft enough to yield to the thumb!

And an orchard is only a family of fruit-trees. Orchards are also very real, but I hope that we do not lose the feeling of the tree. Our affections cling to trees, one by one; and then the orchard becomes almost a sacred spot. A fruit-tree in full load is one of the marvelous objects in nature. We cannot understand how the work is done,—how such abundance is produced and how such color and substance and flavor and faultless form are derived of the crude elements of soil and sunshine and air. It gives of itself out of all proportion to the care and affection that we bestow on it. It is a very sermon in liberality. It is a great thing when the making of orchards spreads rapidly, for it means not only commercial thrift but a growing appreciation of the tender and refreshing products of the earth.

Peach

It is Peach Day, that you have set aside to celebrate the products of this Peninsula. I trust we have come with sentiments alert, with a readiness of adoration of the fruit that is so characteristic of this alluring land between the ocean and the bay.

Here I hold a peach. It is a shapely oblong-spherical body nearly three inches in diameter, pleasant to clasp in the fingers, choice in its fragrance, captivating in its intergrade of tints. I do not know why it came here. I know that last winter a bare tree stood in yonder orchard, giving no sign of any intention but to be a bare tree. Then one day it shook itself loose in the glory of the resurrection we know as spring, and a sheet of pink brilliancy covered it.

The blossoms fell. Leaves came. A little object began to swell on a last year's twig, white-gray and fuzzy and solid. A brown dry papery ring fell from its end. The thread-like point withered and dropped away. The object gradually grew, we do not know why, it became as large as a marble and almost as hard, the white-gray fuzz turned to green, a groove showed along its side. Presently it took form, a blush was on the sunny side, and a passer-by exclaimed, "Oh, there is a peach."

A man from Mars, perhaps one no farther away than the depths of the great city yonder, seeing this savory fruit in my hand and the flexile tree in the orchard, would not connect one with the other.

Out of the tree, bare but a few months ago, this great peach has come, the birth of a twig no thicker than my pencil. Tree and twig and peach all came out of the soil and the air. This peach is oxygen, yet you never saw oxygen to recognize it as a separate substance; it is hydrogen, yet you have not seen hydrogen as an entity; it is carbon, the carbon you see in yonder smoke; it is nitrogen, that you have not perceived as such although you are always within it; it is calcium, magnesium, phosphorus, that you have seen only in their compounds; it is iron, the iron that is in the locomotive even now belching to start from the station over there; it is potassium, and other elements beside.

It is water,—water delicately and deliciously flavored with many intricate compounds. Perhaps this peach is nearly ninety per cent water, yet so nicely is the fluid held in fiber and cell that I revolve the fruit as I may and it does not spill.

This peach is sunshine. It is night, the twilight, and the dawn. It is dew and rain. It is noon, and wind, and weather. It is heat and cold. It is the sequence of the seasons, winter and spring, summer and autumn, and winter again, all of which have gone into the tree that gave it birth.

It is the linkage of the elements and the days, and the showers that freshen the earth. The peach is more, even, than all this: it is a living thing, vital with its own protoplasm, performing a thousand secrets hidden deep in its cells, containing its own energy to assimilate and to grow and to catch the tints of the rainbow and the fragrance of clean fresh winds.

Here with light pressure I part the fruit in halves. The aroma is an elixir. The wrinkled pit or stone is in the center, surrounded by a darker luster like an aureole; for securely inside this stone lies the mysterious kernel, which is an embryo peach-tree; and next year the embryo will not forget to grow, if buried in the ground, nor fail to make a peach-tree; and in the years to come, when you and I shall not be here to see, it or its progeny will bear peaches still.

The continuity of the centuries is in the flat kernel within this stony pit. I do not know why a peach-pit and not a plum-pit is in this place; I do not know whence on the earth the peach came; I do not know how or why this fruit chose or elaborated its nutrients in such proportions as to make itself a peach and not an apricot. Had I before me unlabeled chemical analyses of a peach and an apricot I suppose I could not tell which was which, so nearly would they be alike; size, shape, color, texture of skin and flesh, season, most of the attributes that distinguish the two fruits to us, might not be shown. Yet here is the peach in my hand, perfect and complete; it is mine.

You have made the conditions right. You have chosen the land that the tree might thrive. You have tilled the soil. You have protected the tree from enemies. You have guarded it for several or many years. You have beheld the miracle.

Where There Is No Apple-Tree

The wind is snapping in the bamboos, knocking together the resonant canes and weaving the myriad flexile wreaths above them. The palm heads rustle with a brisk crinkling music. Great ferns stand in the edge of the forest, and giant arums cling their arms about the trunks of trees and rear their dim jacks-in-the-pulpit far in the branches; and in the greater distance I know that green parrots are flying in twos from tree to tree. The plant forms are strange and various, making mosaic of contrasting range of leaf-size and leaf-shape, palm and grass and fern, epiphyte and liana and clumpy mistletoe, of grace and clumsiness and even mis-proportion, a tall thick landscape all mingled into a symmetry of disorder that charms the attention and fascinates the eye.

It is a soft and delicious air wherein I sit. A torrid drowse is in the receding landscape. The people move leisurely, as befits the world where there is no preparation for frost and no urgent need of laborious apparel. There are tardy bullock-carts, unconscious donkeys, and men pushing vehicles. There are odd products and unaccustomed cakes and cookies on little stands by the roadside, where the turbaned vendor sits on the ground unconcernedly.

There are strange fruits in the carts, on the donkeys that move down the hillsides from distant plantations in the heart of the jungle, on the trees by winding road and thatched cottage, in the great crowded markets in the city. I recognize coconuts and mangoes, star-apples and custard-apples and cherimoyas, papayas, guavas, mamones, pomegranates, figs, christophines, and the varied range of citrus fruits. There are also great polished apples in the markets, coming from cooler regions, tied by their stems, good to look at but impossible to relish; and I understand how these people of the tropics think the apple an inferior fruit, so successfully do the poor varieties stop the desire for more. There are vegetables I have never seen before.

I am conscious of a slowly moving landscape with people and birds and beasts of burden and windy vegetation, of prospects in which there are no broad smooth farm fields with fences dividing them, of scenery full of herbage, in which every lineament and action incite me and stimulate my desire for more, of days that end suddenly in the blackness of night.

Yet, somehow, I look forward to the time when I may go to a more accustomed place. Either from long association with other scenes or because of some inexpressible deficiency in this tropic splendor, I am not satisfied even though I am exuberantly entertained. Something I miss. For weeks I wondered what single element I missed most. Out of the numberless associations of childhood and youth and eager manhood it is difficult to choose one that is missed more than another. Yet one day it came over me startlingly that I missed the apple-tree,—the apple-tree, the sheep, and the milch cattle!

The farm home with its commodious house, its greensward, its great barn and soft fields and distant woods, and the apple-tree by the wood-shed; the good home at the end of the village with its sward and shrubbery, and apple roof-tree; the orchard, well kept, trim and apple-green, yielding its wagon-loads of fruits; the old tree on the hillside, in the pasture where generations of men have come and gone and where houses have fallen to decay; the odor of the apples in the cellar in the cold winter night; the feasts around the fireside,—I think all these pictures conjure themselves in my mind to tantalize me of home.

And often in my wanderings I promise myself that when I reach home I shall see the apple-tree as I had never seen it before. Even its bark and its gnarly trunk will hold converse with me, and its first tiny leaves of the budding spring will herald me a welcome. Once again I shall be a youth with the apple-tree, but feeling more than the turbulent affection of transient youth can understand. Life does not seem regular and established when there is no apple-tree in the yard and about the buildings, no orchards blooming in the May and laden in the September, no baskets heaped with the crisp smooth fruits; without all these I am still a foreigner, sojourning in a strange land.

Apple-Year

My last winter apple I ate today.
  Shapely and stout in their modelled skins
  Securely packed in my cellar bins
Two dozen good kinds of apple-spheres lay.

And today I went to my orchard trees
  And picked me the first-ripe yellow fruits
  That hung far out on the swinging shoots
In summer suns and the wonder-day breeze.

And thereby it was that the two years met
  Deep in the heart of the ripe July
  When the wheat was shocked and streams were dry;
And weather of winter stayed with me yet.

For I planted these orchard trees myself
  On hillside slopes that belong to me
  Where visions are wide and winds are free
That all the round year might come to my shelf.

And there on my shelves the white winter through
  Pippin and Newtown, Rambo and Spy,
  Greening and Swaar and Spitzenburg lie
With memories tense of sun and the dew.

They bring the great fields and the fence-rows here,
  The ground-bird's nest and the cow-bell's stroke
  The tent-worm's web and the night-fire's smoke
And smell of the smartweed through all the year.

They bring me the days when the ground was turned,
  When the trees were pruned and tilled and sprayed,
  When the sprouts were cut and grafts were made,
When fields were cleaned and the brush-wood piles burned.

And then the full days of the ripe months call
  For Jefferis, Dyer and Early Joe
  Chenango, Mother, Sweet Bough and Snow
That hold the pith of high summer and fall.

All a-sprightly and tart the crisp flesh breaks
  And the juices run cordial and fine
  Where the odors and acids combine
And lie in the cells till essence awakes.

I taste of the wilds and the blowing rain
  And I taste of the frost and the skies;
  Condensed they lie in the apple guise
And then escape and restore me again.

So every day all the old years end
  And so every day they begin;
  So every day the winds come in
And so every day the twelve-months blend.

V

Spring to Winter

It is a wonderful world in which we live and
vegetation is the garnishment of it. It is a marvelous
experience to see the manifold forms of life emerge
from the bare earth, all the greater because we do not
recognize the sensation. As we are inhabitants of the
planet it surely is our part to appreciate and utilize the
objects and incidents with which we are placed. The
interest in them, and their meaning for us, lies in their
essential nature and in the fact that they are partners,
and the sense-sensations of color and fragrance are
only attributes. The seasons are an integral part of
life; to one who loves the seasons, the garden is the
best personal expression of them.

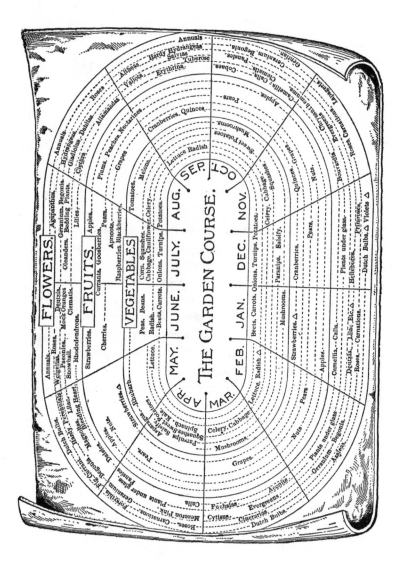

The Garden Flows

Recently I took from its shelf Thomson's "Seasons", to read once more the descriptions and to feel the beneficence of the passing years. This copy was published in my wildering youth as a "school and academic edition", although I do not remember its use in the ungraded school I attended, but the book has had good place in my memory. Here I opened to the first chapter, Spring, and I read

> By Nature's swift and secret working hand
> The garden flows.

Probably more than two hundred years ago James Thomson wrote that strain. It is as fresh and lively this morning as in that far time in another land. The Spring is product of the Winter, as Winter is product of Autumn and Autumn the product of Summer and Summer of Spring. We like to think of Spring, in our northern latitudes, as the real beginning of the year because the plants start to grow and the visible miracles to unfold. Yet the roots or the seed grew last year, which was then the new year, and life was maintained continuously whatever may have been the snowfall or the temperature. There is no ending and no beginning, only stages in a persisting and everlasting process.

Here is a first lesson for the gardener to learn, that he is speeding and, I hope, conserving the processes of Nature and at the same time deriving unexpressed satisfaction in the effort.

It is said there is nothing new under the sun, but the gardener's reaction is as new and fresh to him as if he were indeed the first of all men. Yet there are new things under the sun. The columbines in my garden this Spring have not been seen before, because I planted the seeds of them myself last July from hybridized stock. James Thomson had not seen what I now behold; it is mine, but it issues in gracious continuity from the years that do not return. I find much satisfaction in this partnership with "Nature's swift and secret working hand," and I know that all my successors in times to come may partake in the kinship: the earth is ever prime and new.

As I read again Thomson's "Seasons" I am estopped by the abundant footnotes and explanations, that forever interrupt the text and break it into analysis. I want to read the text for what it says to me, not for what it may mean to the critic. It was written in appreciation of the rural scene, what we in these later days like to call the out-of-doors. Nor do I care much about the supposed merits or deficiencies of the verse; if I do not like it I cease to read.

These gardens we now grow are products of untold garden lovers in untold places and untold generations. Somewhere, sometime, a plant was taken from the wild and set beside a cabin or a cottage. It propagated its kind, but the kind began a process of self-education, expanding to new forms and statures and colors and fragrances. To one or more of the novelties some gardener gave a name, and the progeny of improvement had begun.

So at length we have the Cup-and-Saucer Canterbury bells, the abounding cannas, resplendent roses, rich autumnal dahlias; and new things come to us when strange countries are opened. The garden is not a temporary affair of one year's enthusiasm. My garden was begun more than seventy-five years ago, although my residence is not the same as then; every year, even in far China, it has renewed itself as one continuous and connected emotion. It is better this year because I had it last year. One year builds into the next.

No person may foretell the gardening of the future. The conditions under which human beings live must be important factors. But the future will grow out of the past because plants come out of the past. However great may be the improvement in varieties, we expect a connected, even though an accelerated process, yet there are natural limitations beyond which new vegetable products cannot go and still satisfy the tastes of sensitive minds. Species of plants new to cultivation will be introduced from the wild, and some of the old ones will lose favor and pass out. Methods of soil manipulation and of control of pests are likely to be modified; but the continuing satisfactions must come simply from the growing of plants. Throughout the centuries the garden flows.

The New Year

The spring has come again. The winter has been long and the roads stopped with snow; but today the mercury is at 70 and there are signs of warm thunder; soon the spouts will be filled with April rain. By twos and threes and more the juncoes are in the open fields, and soon will be leaving for their breeding-places. Up from the South the bluebirds have come, acting yet like strangers, flitting singly across the highways. The pussy willows begin to show, and children are gathering them. The winter wheat makes patches of emerald in the distance. Only in the shady places and on the wooded hills does the snow still linger, in long ribbons behind the fences, and in ditches by the roadside where the hastening water runs underneath; there are great patches of it in the farther forests. We can see where the rabbits have gnawed the bushes above the snow line. Soon we shall be looking for arbutus and the first hepaticas; and even now the alders in the swamps are showing their tassels. Before long, cows will be set afield, and horses will be rolling on the ground and snorting their freedom.

The smell of April is here. It is faint and indefinable, yet it is real. It is an effluvium of unbonded brooks, of pastures and bushy fields, drying roads, gathered up by warm winds and whiffed over the earth. Soon the smell of the soil will be freed, that elemental stimulating odor that is unlike every other and that exhilarates today as when the first man turned the soil. It is a creative perfume that suggests teams afield, growing crops, the very essence of the romantic earth. If there were no other criterion by which to distinguish the real farmer, born to the land, I should know him by his response to the smell of the furrow; this redolence will be his incense, it will be an aroma stronger than the balm of pine woods or the wild tang of the sea, it will bring him from the factory and the city and send him into the field with his plow or with any implement that will open the ground and set its fragrance free. It will unlock old memories, grown dim with the rust of years; it will fill him with dreams of flocks on soft pastures and of corn or cotton in long straight rows; it will inspire him with health; it will vision him of summer and harvest, and set him into the determination of spirit that will carry his year to its finish.

Today a few teams are at the plow in warm early fields. Much land was plowed in the autumn and it lies rough and ready for the harrow; soon the harrow will be on it; birds will be watching for provender, and the whole field will be sweet with the blessing of spring. There is no prophecy to compare with a newly plowed field at the end of winter. With all our chemical and physical and biological analysis, this familiar soil is still a mystery, stolid and immotile and yet a new sensitive problem with every new year and a stronger challenge as a man grows calmer and more wise.

January first we set apart in the calendar as the beginning of the year; but to the farmer the year begins with the first furrow, and this epoch coincides more or less roughly, from south to north in this hemisphere, with the spring solstice, when the sun is on the equator. Then the doors begin to be opened and the new year comes in. The farm may be old and the man may be old, but the new year is as instant and fresh as ever. From this threshold, looking outward with a new heart, let us approach the problem of the farm in a troublous time.

The spring is a season of changes other than in the climate. Naturally, new resolutions are made, or old ones are vivified, when vegetation starts again and the birds begin to build, and the animals smell the open fields. The lands are seen with new eyes after the covered sleep of winter. Business affairs are at an epoch in anticipation of a new year and a new deal. On March first or April first interest is due and taxes are to be paid. In some parts of the country, tenants and hired hands move on April first; auctions are held; there is a general change and shift of goods and plans; the main affairs are settled for the year. When the farmer has passed this epoch he takes a new hold and is ready in earnest for his farming; and he goes at it with zest.

The Dandelion

The first warmth of spring brought the dandelions out of the banks and knolls. They were the first proofs that winter was really going, and we began to listen for the blackbirds and swallows. We loved the bright flowers, for they were so many reflections of the warming sun. They soon became more familiar, and invaded the yards. Then they overran the lawns, and we began to despise them. We hated them because we had made up our minds not to have them, not because they were unlovable. In spite of every effort, we could not get rid of them. Then if we must have them, we decided to love them. Where once were weeds are now golden coins scattered in the sun, and bees reveling in color; and we are happy!

A dandelion is shown in the figure to the left on the next page. It is a strange flower, as measured by those which we have already studied. It appears to have a calyx in two parts or series, and a great number of petals. If we look for the pistils and stamens, however, we find that the supposed simple flower is really complex. Let us pull the flower apart and search for the ovary or seed. We find numerous objects like that in the figure on the right. The young seed is evidently at *e*. There are two styles at *d*, and a ring of five anthers at *b*. The dandelion, therefore, must be composed of very many small and perfect flowers.

"Dandelion" and "Floret of a dandelion."

Looking for the floral envelopes, we find a tube, and a long strap-like part running off to *c*. This must be corolla, for the calyx is represented by a ring of soft bristles, *a*. We have, then, a head made up of quadriserial flowers, or florets, as the individual flowers may be called. The entire head is reinforced by an involucre, in much the method in which the dogwood is subtended by four petal-like bracts and the calla spadix by a corolla-like spathe.

One cloudy morning the dandelions had vanished. A search in the grass revealed numbers of buds, but no blossoms. Then an hour or two of sunshine brought them out, and we learned that flowers often behave differently at different times of the day and in various kinds of weather.

In spite of the most persistent work with the lawn mower, the dandelions went to seed profusely. At first, we cut off many of the flower-heads, but as the season advanced they seemed to escape us. They bent their stems upon the ground and raised their heads as high as possible and yet not fall victims to the machine; and presently they shot up their long soft stems and scattered their tiny balloons to the wind, and when the lawn-mower passed, they were either ripe or too high to be caught by the machine.

This seed has behaved strangely in the meantime. The fringe of pappus (as the bristle-like calyx is called) is raised above the seed by a short, narrow neck (*e*, in the figure on the previous page), when the plant is in flower; but at seed-time this neck has grown an inch long (in the figure on the next page), the anthers, styles and corolla have perished, the pappus has grown into a spreading parachute, and the ovary has elongated into a hard, seed-like body. Each one of us has blown the tiny balloons from the white receptacle, and has watched them float away to settle point downwards in the cool grass; but perhaps we had not always associated these balloon voyages with the planting of the dandelion.

The dandelion, then, has many curious habits. It belongs to the great class of compositous (or compound) flowers, which, with various forms, comprises about one-tenth of all the flowering plants of the earth. The structure of these plants is so peculiar that a few technical terms must be used to describe them. The entire "flower" is really a head, composed of florets, and surrounded by an involucre. These florets are borne upon a so-called receptacle. The plume-like down upon the seeds is the pappus. The anthers are said to be syngenesious ("in a ring"), because united in a tube about the style; and this structure is the most characteristic feature of compositous flowers,—more designative of them, in fact, than the involucrate head, for in some other kinds of plants the flowers are in such heads, and in some compositous flowers the florets are reduced to two or three, or even to one!

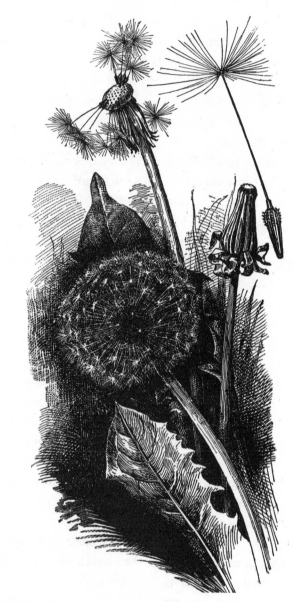

"The dandelion."

The Apple-Tree in the Landscape

The April sun is soft on the broad open fenced fields, waking them gently from the long deep sleep of winter. Little rills are running full. The grass is newly coolly green. Fresh sprouts are in the sod. By copse and highway the shad-bushes salute with their handkerchiefs. Apple-trees show tips of verdure. It is good to see the early greens of changing spring. It is good to look abroad on an apple-tree landscape.

As to its vegetation, the landscape is low and flat, not tall. There is a vast uniformity in plant forms, a subdued and constrained humility. A month later the leafage will be in glory, but that also will have an aspect of sameness and moderation. Perhaps the actual variety of species will be greater than in many parts of the abounding tropics, and to the careful observer the luxuriance will be as great, although not so big; but as I look abroad I am impressed with the economy of the prospect. It comes nearer to my powers of assimilation, quiets me with a deep satisfaction; the contrasts are subdued, the processes grade into each other imperceptibly in the land of the lingering twilight.

In this prospect are maples and elms and apple-trees. The maples and elms are of the fields and roadsides. The apple-trees are of human habitations and human labor; they cluster about the buildings, or stand guard at a gate; they are in plantations made by hands. As I see them again, I wonder whether any other plant is so characteristically a home-tree.

So is the apple-tree, even when full grown, within the reach of children. It can be climbed. Little swings are hung from the branches. Its shade is low and familiar. It bestows its fruit liberally to all alike.

The apple is a sturdy tree. Short of trunk and short of continuous limb, it is yet a stout and rugged object, the indirectness of its branching branches adding to its picturesque quality. It is a tree of good structure. Although its limbs eventually arch to the ground, if left to themselves, they yet have great strength. The angularity of the branching, the frequent forking, the big healing or hollow knots with rounding callus-lips, give the tree character. Anywhere it would be a marked tree, unlike any other.

The bark on the older surface sheds in short oblong irregular scales or plates that detach perhaps at both ends and often at the sides, clinging by the middle until the curl loosens them and they fall to the ground. These plates or chips are more or less rowed up and down the trunk and on the larger branches, yet the apple bark is not ridged and furrowed as on the elm. The bark is not checked in squares as on old pear-trees nor peeling as on cherries. In dry weather, the loose old bark is dark brown-gray, often supporting gray lichens, but in rain it is soft and nearly black, yielding pleasantly to the touch. In the forks, the bark is not so readily cast and there the chips may lie in heaps. On the young limbs and small trunks the bark is tight and close, not splitting into seams or furrows with the expansion of the cylinder but stretching and throwing off detached flakes and chips. Under the chips various insects hide or make some of their transformations. There the codlin[1] moth pupates. The old remains of scale insects may be found on the exterior. In the furrows about the dormant buds the eggs of plant-lice pass the winter.

To destroy these breeding and hiding places, many careful apple-growers scrape away the loose bark, being careful not to expose the quick living tissue; and on the younger wood the eggs of aphis and other pests, as well as cocoons and nymphs, are destroyed by vigorous winter spraying. The regular spraying of apple-trees, in the different seasons, more or less sterilizes the bark. Many forms of canker, due to fungi and bacteria, invade the bark, making sunken areas and scars, often so serious as to destroy the tree. All these features are discoverable in the apple-tree.

The trunk of the apple-tree is short and stout, usually not perfectly cylindrical and not prominently buttressed at the base. In old trees it is usually ribbed or ridged, sometimes tortuous with spiral-like grooves, often showing the bulge where the graft was set. The wood is fine-grained and of good color, and lends itself well to certain kinds of cabinet work and to the turning-lathe for household objects; it should be better known.

If left to itself, the tree branches near the ground, making many strong secondary scaffold trunks; but the plant does not habitually have more than one bole, even though it may branch from the very base; it is a real tree, even though small, and not a huge shrub. In the natural condition, the trunk often rises only a foot or two before it is lost in the branches; at other times it may be four or six feet high. Under cultivation, the lowest branches are usually removed when the tree begins to grow, and an evident clean trunk is produced. In Europe and the Eastern States, it has been the practice to trim the trunk clean to the height of four or six feet; but in hotter and drier regions the trunk is kept short to insure against sun-scald; and with the better tillage implements of the present day it may not be necessary to train the heads so high.

In old hill pastures, in many parts of the North, one sees curious umbrella forms and other shapes of apple-trees, due to browsing by cattle. A little tree gets a start in the pasture. When cattle are turned in, they browse the tender terminal growth. The plant spreads at the base, in a horizontal direction. With the repeated browsing on top, the tree becomes a dense conical mound. Eventually, the leader may get a strong headway, and grows beyond the reach of the browsers. As it rises out of grasp, it sends off its side shoots, forming a head. The cattle browse the under side of this head, as far as they are able to reach, causing the tree to assume a grotesque hour-glass shape, flat on the under part of the head, with a cone of green herbage at the ground. Sometimes pastures are full of little hummocks of trees that have not yet been able to overtop the grazers.

The winter apple-tree in the free is a reassuring object. It has none of the sleekness of many horticultural forms, nor the fragility of peaches, sour cherries and plums. It stands boldly against the sky, with its elbows at all angles and its scaly bark holding the snow. Against evergreens it shows its ruggedness specially well. It presents forms to attract the artist. Even when gnarly and broken, it does not convey an impression of decrepitude and decay but rather of a hardy old character bearing his burdens. In every winter landscape I look instinctively for the apple-tree.

We are so accustomed to the apple-tree as a part of an orchard, where it is trimmed into shape and its bolder irregularities controlled, that we do not think it has beauty when left to itself to grow as it will. An apple-tree that takes its own course, as does a pine-tree or an oak, is looked on as unkempt and unprofitable and as a sorry object in the landscape, advertizing the neglect of the owner. Yet if the apple-tree had never borne good fruit, we should plant it for its bloom and its picturesqueness as we plant a haw-thorn or a locust-tree.

In winter and in summer, and in the months between, my apple-tree is a great fact. It is a character in the population of my scenery, standing for certain human emotions. The tree is a living thing, not merely a something that bears apples.

from *Lessons of To-day*

Above all, old and young, we must never lose faith in the soil.
It is the source and condition of our existence. It never grows
stale and it never wears out. The earth is always young.

The fields were parched with summer's heat,
    The life and green from swamps had fled,
The dry grass crunched beneath the feet,
    And August leaves dropped stiff and dead.

Then light south winds 'cross wood and shore
    Brought cooling clouds and slow sweet rain,
And hills and crops were new once more
    And grasses greened on marsh and plain.

So swift the magic sent its spell
    Thro' burning corn and pastures dumb,
'Twas clear the world had rested well
    Against the time when rain should come.

So virile is this earth we own
    So quick with life its soil is stung,
A million years have come and flown
    And still it rises green and young.

Leaves

A gain the leaves are falling. They cover the ground, but the twigs are bare. The leaves will lie in drifts and windrows; by spring they will be crunched and broken, and by autumn they will have passed into the forest floor and will be leaves no more. The twigs will stand all winter in the bitter air, apparently mere lifeless things; but one day in April or pregnant May the tinge of life will be in them, buds will swell and burst and soft new fabrics will unfold and spread themselves to the warming sun. How they will come or why no man is wise enough exactly to explain.

The twig that this year bore an ovate leaf will next year bear the same, with veins and margins of a pattern formed in some remote dim past, with minute hair and coverings of their own kind and theirs alone. And other twigs will bear leaves of other forms and patterns, likewise all their own, as unerringly as the Earth passes from Libra into Taurus, as steadfastly as the sun comes back to the equinox of March. So steadfast are they of their process and their destiny that by them men may prophesy.

A flat thin fragile tissue is a leaf, that detaches itself at the appointed season, keeping good time of the procession of the year. Why it should come from a round hard twig is beyond all calculation. Seen apart and without experience of miracles, no man would think a leaf and twig of the same genesis. They would not suggest relationship one with the other. And still less would they suggest mothership of the inert ground, or the transparencies of sunshine and of air. At these marvels men should stand with reverent mein, as in the midst of mysteries.

The leaves are falling one by one. Miracles are in the air.

The Garden of Gourds

It is the first day of October. Summer birds are gone. Frosts have signaled for winter. All tender vegetables are harvested from the garden. Marigolds and zinnias and a few other rugged things still retain the glow of warm weather, yet autumn chill is in the air, new colors are on the hills, dead leaves begin to cover the grass. It is plain we approach a great event in the progress of the year, when products and ambitions will be housed and we shall settle down to the hopeful routines of winter. Most persons will cease to regard the landscape and they will seldom go to the garden or follow the lines of the brook.

Yet even now the gourds hang on the trellises, and although blossoms are mostly gone and leaves have passed their prime, the bright green-and-yellow striped fruits in comely attractive form ask for attention. Probably in every year since my youth—and that was long ago—I have grown gourds of one kind or another and sometimes of many kinds. I have made an herbarium collection of them, for record, including foliage and flowers; and many fruits lie in boxes. I cannot remember when I did not know them. For time beyond recollection I have wanted to write a simple book about the gourds; my technical and scientific writings on the Cucurbits began nearly fifty years ago; and now, as the years are ripe, I wish to express my joy in the experiences.

Once the gourds were common objects in homes, then other blooms and fruits displaced them, and now they are back again in a new vogue of popular favor, but for myself the interest in them has never lagged. They are so shapely and so colorful, so strange in their markings, so endlessly unlike each other, so durable in winter months, so apparently unrelated to the vines that bear them, and yet so simple to grow, that they hold the interest tenaciously.

The present interest in gourds is well attested by the organizing of the International Gourd Society, headquartering in California. Semi-annually it publishes a Gourd Bulletin.

THESE GOURDS THAT hang so handsomely on their vines today are of the yellow-flowered kind, from two of which the colored frontispiece of this book[2] is painted. Seeds of them were planted in the open ground on June 12th, 1936, which is about three weeks later than usual practice for Ithaca, New York. The season was dry and hot; no special attention was given except that the soil was warm mellow and fertile, and exposure was to full sun. In ten days to two weeks the seeds had begun to germinate. In about a month the plants had started to make four to six small true leaves but showed no sign of running and bore no flowers. In early August the plants were beginning to climb and a few staminate or male flowers had appeared. The middle of September the vines were all over the wire trellis and colorful gourds were in the wind.

Advantage of the gourds as garden plants is the fact that in North America they require no fussing care. Persons who drench and sprinkle them may see the fruits fail or drop and perhaps have mildew spread over the vines. Apparently the plants this year bore as profusely and about as early as those customarily started late in May. Gourds are heat-loving plants and they do not profit by being so far advanced that "slow" weather overtakes them. At both ends of the season they are killed by frost. They are warm season plants. They are annuals; or if some of them are perennial, they are treated as if annual in northern regions.

The plants perish with the first good frosts of autumn; then the old vines are pulled from the trellises or out of the bushes, all the ripe hard gourds are harvested if not already picked, and the brilliant shapely fruits carry the three summer months through to the holidays and beyond. One is impressed that here is real harvest, and it is distinctly unlike the usual autumnal produce and flowers.

Gourds are excellent subjects for home decoration as well as for exhibitions in schools and fairs. Before me is the program of a flower-show in which an "educational gourd exhibition" is a feature. This program covers two pages, including correlations with art, nature-study, literature and handicrafts. There is wide opportunity, in such an assembly of material, for design and color effects. Often the gourds, when not naturally colored (as the Lagenarias) are painted and decorated by hand, the inside having been removed and the shell thoroughly dried and cured. Sometimes the gourds are partially carved or trimmed.

On the subject of arrangement, Helen M. Tillinghast writes in her "First Gourd Book" as follows: "Pottery, old pewter plates, copper trays, baskets, wooden chopping bowls can be used for containers. As to grouping, it is well to remember that a smooth gourd emphasizes the charm of a warted one, a white or plain yellow relieves the monotony of too many stripes or splashes of different markings."

In the residence, baskets of gourds of different shapes and colors are striking articles of ornament, never failing to interest visitors. They may be used singly in plans of decoration. Florists make up clusters of them in hanging designs, using nuts and other autumnal objects among them.

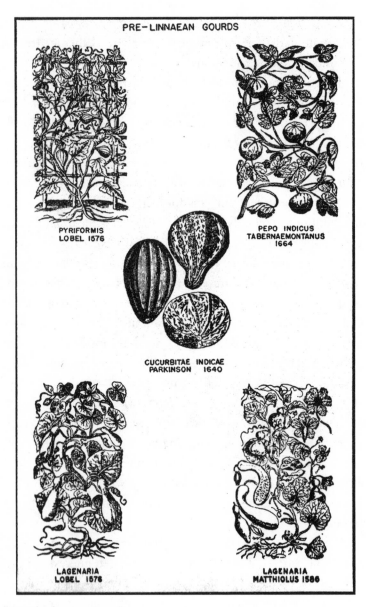

"Gourds of centuries ago."

Colors of the brilliant yellow-flowered gourds are not permanent although they last well until Christmas or beyond as a rule. The gourds should not be harvested until ripe and fully colored, when the shell is hard, and then they are dried. If stored in a closely confined space or in a moist place they are likely to decay or at least to develop sunken moldy spots. As I read the proofs on this book in the early part of January, 1937, a goodly company of gourds is before me, the colors still bright and the varying shapes most attractive.

In preparation for decorative uses the gourds should be cleaned and softly polished, then rubbed with a good floor wax. Some users varnish them, but this adds an unnatural and too prominent gloss. The fruits should be dry when wax or other covering is applied.

Yet, the primary interest in gourds should be that of the gardener and the horticulturist. They may have a definite place, associated with seasons and birds and gardening labor. The germination will be interesting, with the way in which they pull themselves out of the seed-coat hood and then straighten themselves to the sky; the rapid continuous growth is reassuring; blossoms of the two separate kinds are stimulating; and there is never-ceasing satisfaction in the forthright development of the fruits. It is usually my practice to grow the different kinds in separate parts of the garden as far as possible, and I find myself going from one planting to the other to make comparisons in progress; and although I have known them for so many seasons I never cease to wonder at the shapes and colors.

Lesson I.—The Pumpkin

"Plant of Cucurbita Pepo."

O,—fruit loved of boyhood: The old days recalling,
When wood-grapes were purpling and brown nuts were falling!
When wild, ugly faces we carved on its skin,
Glaring out through the dark with a candle within
When we laughed round the corn-heap with hearts all in tune,
Our chair a broad pumpkin,—our lantern the moon,
Telling tales of the fairy who travelled like steam,
In a pumpkin-shell coach with two rats for her team.

John Greenleaf Whittier

In October the cornfields were golden with pumpkins. The corn
was in shocks. The tassels were ripe and dry, and hung down-
ward as if mourning for the dying year. The maple leaves, yellow

and red, were falling to the ground like flocks of brilliant birds. Lonely hickory trees held on to their dun-yellow leaves as if loth to let them go. But the pumpkins seemed to be in their prime. Fat and sleek they lay between the corn shocks, and shone out among the drying weeds. We did not remember to have seen them before.

It is now November. Heavy frosts have come. One night the brook was frozen nearly to its middle. Much of the corn is still in the shock, but the pumpkins have been taken under cover. They lie in heaps on the barn floor. The hay and straw falls over them. Still the old cow can smell them. I like to sit on them, and run my fingers down their smooth, broad grooves.

In some parts of the State, I miss the pumpkins in the cornfields. These are the regions in which there are many silos; corn is grown in large fields; corn harvesters are used. The absence of the pumpkin tells me of a change in the kind of farming since I was a child. Now you want to ask me some questions; but I shall ask them first of you. Some one in school or in your neighborhood can answer them for you, if you do not know.

My first questions are these:

1. What is a silo and what is it used for?
2. What do the farmers who have silos raise to sell?
3. Why are pumpkins so often planted amongst corn?
4. Do you know of any other kinds of plants that grow underneath taller plants (look under the trees, even in the dooryard; and in the orchard; and under the currant bushes)?

But you must tell me something about the pumpkin itself. You have made jack-lanterns, and you will know. You know the Hubbard squash; but why did you never make a jack-lantern from one? Look at the pumpkin and the Hubbard squash again. If you live in town, you can stop at the grocer's and see them. But I suggest that you children take some pumpkins and squashes to the schoolhouse,—all the kinds you can find. Put them on the platform or on the table. If they are nicely arranged, I am sure you will think them handsome. Then write Uncle John.

5. How many different kinds (or shapes) of pumpkins do you know?
6. What kind of stem does the pumpkin have? How does it differ in this respect from the Hubbard Squash?
7. Look at the blossom end of both; how do they differ?
8. Look at their shape and tell how they differ.
9. Can you see any difference in the seeds of the pumpkin and the Hubbard squash?
10. Explain any difference in color. If you cannot secure a Hubbard squash, then make the observations on the pumpkin alone.

November: June

The frost is here again. It has blasted the tomato vines and beans; the cucumber shoots are limp with blackened withered leaves; the stately rows of sugar corn rustle dryly in the wind; the last cosmos and dahlia are gone, and the proud bushes that bore the flaring blooms are broken and dead; the China asters and the marigold are in ruins.

So has the garden gone; the hopes of June with the achievements of August and September are passed again into the burdened years. A tinge of sadness is in the crisp autumn air, the low sun is only mildly warm at noon, and twilight creeps on before the day's work is done. Here is the wreck of the year; all the energies that burst in April are spent, the leaves loose their hold in a million appointed places and fall aimlessly into unassorted heaps. One would think that defeat and death are everywhere. The deadness of the winter night is even yet marching on the landscape. It is accounted a sad and ineffective ending for the brilliant promises of May.

A squirrel is harvesting the fallen nuts, to store them against the needs of winter; he is alert and quick, and apparently has neither fear nor dread. The hens go in and out contentedly. Cattle are in peace on distant fields. Crows flap out and away at dawn and back again at night as they flew to feeding-grounds in summer.

The pinks that spiced the air in May and June are making bright aspiring shoots from the bottoms of their grassy clumps: another May is coming. Young hollyhocks with great bold leaves are along the margins: another July is in their roots. Foxgloves, snapdragons, and campanulas are sprouting at the base in sheer anticipation. Fresh tufts have sprung where the Madonna lilies held their stately bloom when June was passing to July. Great buds are on the crowns of the Christmas rose, to give bloom again before St. Patrick's Day, and the rhubarb crowns are ready. The twigs of trees from which the dead leaves fell are heavy with great buds that have harvested the vigor of summer and that will burst in leaf and bloom when the swallows come again. Men are plowing in the fields, to make ready for a new earth. Fences are building, old accumulations of worn-out things are burning in their heaps, highways are receiving the final betterments, crops are in the harvest. Men and pinks and squirrels are instinct with ancient faith.

It has been a brilliant year, when beast and tree and men have ridden one more journey around the sun and have come back with a harvest. The holy harvest is in the season's brood, the showers of fallen golden leaves, the preparation for another May, the year's accomplishment and the ripened soul. The energies of June are garnered in November.

An Outlook on Winter

In the bottom of the valley is a brook that saunters between ooz-ing banks. It falls over stones and dips under fences. It marks an open place on the face of the earth, and the trees and soft herbs bend their branches into the sunlight. The hang-bird swings her nest over it. Mossy logs are crumbling into it. There are still pools where the minnows play. The brook runs away and away into the forest. As a boy I explored it but never found its source. It came somewhere from the Beyond and its name was Mystery.

The mystery of this brook was its changing moods. It had its own way of recording the passing of the weeks and months. I re-member never to have seen it twice in the same mood, nor to have got the same lesson from it on two successive days; yet, with all its variety, it always left that same feeling of mystery and that same vague longing to follow to its source and to know the great world that I was sure must lie beyond. I felt that the brook was greater and wiser than I. It became my teacher. I wondered how it knew when March came, and why its round of life recurred so regularly with the returning seasons. I remember that I was anxious for the spring to come, that I might see it again. I longed for the earthy smell when the snow settled away and left bare brown margins along its banks. I watched for the suckers that came up from the river to spawn. I made a note when the first frog peeped. I waited for the unfolding spray to soften the bare trunks. I watched the greening of the banks and looked eagerly for the bluebird when I heard his curling note somewhere high in the air.

Yet, with all my familiarity with this brook, I did not know it in the winter. Its pathway up into the winter woods was as unexplored as the arctic regions. Somehow, it was not a brook in the winter time. It was merely a dreary waste, as cold and as forbidding as death. The winter was only a season of waiting, and spring was always late.

Many years have come and gone since then. My affection for the brook gave way to a study of plants and animals and stones. For years I was absorbed in phenomena. But now mere phenomena and materials have slipped into a secondary place, and the old boyhood slowly reasserts itself. I am sure that I know the brook the better because I know more about the things that live in its little world; yet that same mystery pervades it and there is that same longing for the things that lie beyond. I remember that in the old days I did not mind the rain and the sleet when visiting the brook. I was not conscious that they were not a part of the brook itself. It was only when I began to dress up that the rain annoyed me. I must make a proper appearance before the world. From that time, the brook and I grew farther apart. We are coming together again now. It is no misdemeanor to get wet if you feel that you are not spoiling your clothing. One's happiness is largely a question of clothes.

But the brook is one degree the better now just because it remains a brook all winter. The winter is the best season of the four because there is more mystery in it. There is a new and strange spirit in the air. There are strange bird-calls in the depths of the still white woods. There are strange marks in the new-fallen snow.

There are soft noises when the snow drops from the trees. There are grotesque figures on the old fence. There is the warm brown pathway of the brook still winding up between oozing banks. In the spring there are troops of flower-gatherers along the brook. In the summer there are fishers at the deep pools. In the fall there are nut-gatherers and aimless wanderers. In the winter the brook and I are alone. We know.

Most of us, I fear, look upon winter with some feeling of dread and apprehension. It is to be endured. This feeling is partly due to the immense change that comes with the approach of winter. The trees are bare. The leaves are drifting into the fence-rows. The birds have flown. The deserted country roads stretch away into leaden skies. The lines of the landscape become hard and sharp. Gusty winds scurry over the fields. It is the turn of the year.

To many persons, however, the dread of winter, or the lack of enjoyment in it, is a question of weather. We speak of bad weather, as if weather ever could be bad. Weather is not a human institution, and is not to be measured by human standards. There is strength and mighty uplift in the roaring winds that go roistering over the winter hills. The cold and the storm are a part of winter, as the warmth and the soft rain are a part of summer. Persons who find happiness in the out-of-doors only in what we call pleasant weather have not found the great joys of the open fields.

We speak of winter as bare, but this is only a contrast with summer. In the summer all things are familiar and close; the depths are covered. The view is restricted. We see things near by. In the winter things are uncovered. Old objects have new forms. There are new curves in the roadway through the forest. There are steeper undulations in the footpath. Even when the snow lies deep on the earth, the ground-line carries the eye into strange distances. You look far down into the heart of the woods. You feel the strength and resoluteness of the framework of the trees. You see the corners and angles of the rocks. You discover the trail that was lost in the summer. You look clear through the weedy tangle. You find new knot-holes in the tree-trunks. You penetrate to the very depths. You analyze, and gain insight.

Many times in warm countries I have been told that the climate has transcendent merit because there is no winter. But to me this lack is its disadvantage. There are things to see, things to do, things to think about in the winter as in the spring. There is interest in the winter wayside, in the hibernating insects, in the few hardy birds, and the deserted nests, in the fret-work of the weeds against the snow, in the strong outlines of the trees, in the snow-shapes, in the cold deep sky. To many persons these strong alternations of the seasons emphasize and punctuate the life. They are the mountains and the valleys. The winter is a part of the naturalist's year.

The lesson is that our interest in the out-of-doors should be a perennial current that overflows from the fountain that lies deep within us. This interest is colored and modified by every passing season, but fundamentally it is beyond time and place. Winter or no winter, it matters not: the fields lie beyond.

Midwinter

The November days are running hard into December. The nights are long; and the glow of my study light is like an island in the night, in the night that is dark and deep and heavy with winds and the driven leaves. The "dead of winter" comes on. The days will be narrowed to their smallest straits, and the darkness will dominate. Only a few drifting window birds are in these days, only seldom is there a wild four-foot, the insects and the lesser world are tucked away so far and so compactly that they might never have been.

Soon will the jingle bells give voice. The white-lights along every street will gleam in all their glory. Every shop will be resplendent. The giddy whirl inside will go round and round. How expediently do we try to make up for the dead of winter!

Yet now is the world revealed. The mask is stript. The way it sleeps, the fashion of it,—this I want to know. How the tree looks in its dormancy, where are the cocoons, how the frost comes to the creek, where the rosettes of many weeds hug the ground and how the vegetation shrinks into the cold, the way of the blowing winds, how the snow comes down,—these I want to know. One world from December unto December, with its wonder shifts, one nature preparing and rising and sleeping and then preparing again forever and forever: this is the great joy of it,—not the waiting for a new season, but the wonder-panorama of the present! This is Jackman's "rolling year."

To be a part of it, to expand in the blessed uncovered days of winter, to feel the leap of spring (what means this word "spring"), to burn with the summer, to ripen with the autumn, and then again to go down into the wild winter: verily, this is life!

Greenhouse in the Snow

"Who loves a garden loves a greenhouse too," wrote Cowper in "The Task." The greenhouse of a century and a half ago was a very different affair from the glasshouses of the present day both in construction and in the contents, yet then as now it was "warm and snug, while the winds whistle, and the snows descend." It was in this vein that I wrote an article on the greenhouse in the snow many years ago for Country Life in America when editor of that new magazine. I had not then read the poet Cowper's meditations on these subjects, else we might have foregone the essay; but as the article expresses experiences that we are likely to lose with the smaller interest in the "unobtrusive greenhouse as an adjunct to a modest home" it is reprinted here.

We fear that the garage is taking the place of the greenhouse, and that the sensation of rapid movement has a stronger appeal than the meditative quiet of handwork in a little glass house. We deplore the passing of this kind of greenhouse with the interesting range of plants and appliances that went with it, and we have spoken of the subject in the first essay in this book.[3] One is never in such intimate contact with plants as in a greenhouse. The effort is all concentrated in this little spot. The pots may be shifted at will. Every shoot may be trimmed or trained. Insects are under better control than elsewhere; at all events we can make the climate noxious to them by the various processes of fumigation. Tender plants may be carried over from year to year. Plants strange to the region find congenial home.

When the night has lengthened so that it is dark at supper time, when the leaves are coloring on the fields, when frosts threaten and the summer growths are near their end, there is no satisfaction more keen than to prepare for the protected garden under glass. Summer plants that are still vigorous, as petunias and snap-dragon, may be lifted for the greenhouse; late seedlings of calendulas and schizanthus are brought inside; paper-whites, freesias, anemones and tuberous crowfoots are planted in pots or flats for winter and spring bloom, each kind receiving its requisite treatment; carnations are brought in. If the greenhouse garden is a year-round enterprise, attention is now centered in it with new emphasis and all the plants look better in the contrast to the bleakness of the landscape. As the verdure goes down in the garden other greenery extends itself under glass. We begin to feel that exclusive sensation of cold climates, the gratification of the housing instinct.

For the practical operations of gardening, an unheated greenhouse is a useful adjunct, and in it one may feel something of the sensation peculiar to a greenhouse proper. A house eight or nine feet wide and eighteen or twenty feet long, made of large hotbed sash, answers the purpose very well, allowing a walk below the peak and a good bench or bed on either side. The season may be lengthened a month at either end, and if a little hand heater is employed many plants can be carried until the hard weather of winter sets in. It is surprising how many plants of divers kinds can be handled in such a house; but it would hardly tempt me to write my sentiments about the greenhouse in the snow.

In the warmer parts of the United States the lathhouse is an effective structure (although not necessarily made of lath), protecting from cold in the winter and excessive heat and sun in the summer. It is capable of being made an attractive architectural feature of a residence. I think it will occupy an important place in gardening in years to come, even in cold regions for wind and summer as well as winter protection.

The plants of Cowper's greenhouse would hardly be deemed important for the purpose with us, and he mentions few. The "spiry myrtle," orange, lime, amomum by which pimenta may have been meant, geranium with "crimson honors," ficoides or mesembryanthemum, are listed. Probably the greenhouse Cowper knew was the early structure in which, as the name indicates, plants were to be kept green through the winter rather than a hot-house or growing-house. The "greenhouse" of our day is a highly developed and effective structure in which climate and environment may be controlled to a nicety and a novel kind of ecology may be established. But we set out to introduce the article of twenty-five years ago, and now it follows only slightly modified.

It is in the dead of winter that the greenhouse is at its best, for then is the contrast of life and death the greatest. Just beyond the living tender leaf—separated only by the slender film of the pane—is the whiteness and silence of the midwinter. You stand under the arching roof and look away into the bare blue depths where only stars hang their cold faint lights. The bald outlines of an overhanging tree are projected against the sky with the sharpness of the figures of cut glass. Branches creak and snap as they move stiffly in the wind. White drifts show against the panes. Icicles glisten from the gutters. Bits of ice are hurled from trees and cornice, and they crinkle and tinkle over the frozen snow. In the short sharp days the fences protrude from a waste of drift and riffle, and the dead fretwork of weed-stems suggests a long-lost summer. There, a finger's breadth away, the temperature is far below zero; here, is the warmth and snugness of a nook of tropic summer.

This is the transcendent merit of a greenhouse,—the sense of mastery over the forces of nature. It is an oasis in one's life as well as in the winter. One has dominion.

But this dominion does not stop with the mere satisfaction of a consciousness of power. These tender things, with all their living processes in root and stem and leaf, are dependent wholly on you for their very existence. One minute of carelessness or neglect and all their loveliness collapses in the blackness of death. How often have we seen the farmer pay a visit to the stable at bedtime to see that the animals are snug and warm for the night, stroking each confiding face as it raised at his approach! And how often have we seen the same affectionate care of the gardener who stroked his plants and tenderly turned and shifted the pots, when the night wind hurled the frost against the panes! It is worth the while to have a place for the affection of things that are not human.

Did my reader ever care for a greenhouse in a northern winter? Has he smelled the warm, moist earth when the windows are covered with frost? Has he watched the tiny sprout grow and unfold into leaf and flower? Has he thrust a fragment of the luxuriance of August into the very teeth of January? There is no place where the drenching cold rain sounds so good and where one feels so neatly housed from the storm as under a greenhouse roof; it is pleasant to see the streams run down the glass and leap from the eaves.

Greenhouses are of many kinds. There is one kind of the commercial plantsman, and another of the man of means whose conservatory is essential to the architectural completeness of his mansion. Of these we need not speak here, for their necessity is long ago established. But for another kind we wish to plead,—for the quiet unobtrusive greenhouse as adjunct to a modest home.

The object of this simple winter garden need not be the mere growing of flowers, although these may be had without trouble. It is worth the while to grow a plant just because it is a plant and because we are human beings. The best plant is the one that has the deepest significance to you, even though it never makes a flower. I know a man who has hundreds of plants in expensive greenhouses, and the best plant of all is a little white clover that closes its leaves by night and opens them by day.

Against the background of winter every green and growing plant is emphatic. Against the luxuriant background of summer, a plant twice as good may be overlooked and lost. The simplest and easiest things are best, for it is not well to make the uncommon things too common. A dainty rarity is all the better because it is seen in contrast with the homespun of the geranium and begonia; and the common things perpetuate the continuities and purposiveness of our lives.

Like all effort that is worth the while, the labor of growing plants under glass requires watchful care. This care is its own reward. Many plants, however, are easy to grow, and with these the novice should begin; and with them, also, the very busy man should be content. All of us can grow bulbs. We can lift the roots of petunias and alyssum from the garden when the frost comes. We can start the seeds of many annuals in late summer. We can make cuttings of begonias and coleus and a score of common things. Here and there we can pick up something new. Gradually we add to our store; and in three years' time our winter garden, small or large, becomes a unique collection of old-time friends and of new-time rarities.

The Garden of Pinks

From earliest boyhood the pinks have been my companions.
Mounds and rings of Grass pinks were in the front yard, left
there by my mother, so different in their delicacy from the weeds
and brush and deep smells of the forest from which the farm was
cut that they seemed like tokens from another and remoter earth.
Their fresh colors and spicy fragrance were of a different order of
things, and they led me out to hopes of far countries. To this late
day the memory of them lingers.

Why we called them Grass pinks is unknown to me, whether
from their grass-like foliage or because the yard-grass, with reck-
less impertinence, crowded them out. I liked the name, because
the grass that succeeded the old woods seemed to establish us as
part of the world of people and houses; yet it is not a good name,
seeing that all pinks are grass-like and it does not suggest the other
qualities of the plants. I later learned that the quotable herbalist,
John Gerard, was likewise puzzled what to call them, so much so
that in his volume of 1597 he wrote of "faire double purple flowers
of a sweete and spicie smell, consisting of five leaves, sometimes
more, cut or deeply jagged on the edges, resembling a feather;
whereupon I gave it the name Plumarius, or feathered pinke". This
is a name for tomes and those who persist in writing them; but to
me the plant is Cottage pink, and so I call it in this my book.

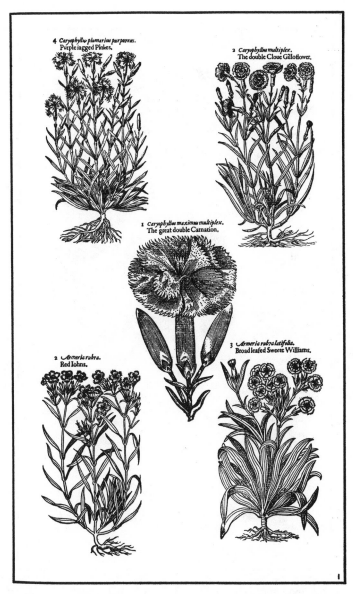

"Pinks of John Gerard, 1597, representing, in order, purple jagged pinks, the choice clove gilloflower, the great double carnation, Red Johns, broad leafed Sweete Williams."

"Pinks of James Vick, 1872. Top row: left, Picotee; right, Double Pink. Middle row: left, Dianthus laciniatus flore-pleno; right, Dianthus diadematus flore-pleno. Bottom row: left, carnation; right, Dianthus Heddewigii. All as named by Vick."

It was in my early years that *James Vick's Catalogue and Floral Guide* found its way from Rochester, New York, to our home, and this disclosed abundant wonders and offered seeds of them. His Catalogue for 1872 is in front of me as I write, being then the tenth annual issue, and I turn first to Dianthus, "a splendid genus of the most beautiful perennials grown". It was probably from Vick that I learned of the Chinese pink and the carnation and how to grow them, although in those days the carnation was not a glasshouse winter flower. These three I now know as *Dianthus plumarius, D. chinensis,* and *D. caryophyllus;* but the second one still has no satisfactory English name and therefore in this book I call it Rainbow pink. The name Carnation, for the third one, is no longer appropriate, inasmuch as the flowers may be of any color other than carnation and reddish, but we will let that pass.

These former pinks are reproduced from original prints in the two preceding figures, in reduced size.

For any number of years I have intended to write a book about pinks, both to express my joy in them and to clarify the seemingly hopeless confusion into which the names and kinds have fallen, as well, also, as to stimulate a renewed interest in these desirable plants. Ten years ago I wrote a brief essay about them for my little book, The Garden Lover; and as I have never heard of anyone who has read that book I reprint the article here, with only trivial alterations, after which we may proceed to more important matters.

It is the first day of the year. Two months and more ago the leaves were fallen; the bushes and the ground-covers in the woodlands have gradually lost their greens and purples and have become russet-brown. Yet the browns are not today the tones of the landscapes, for the snow has sifted through the forest and settled on the fields. The old accustomed walks into the woods are one with the broad undefined spaces; the trilliums are sleeping somewhere; the birds of summer are in far lands, where I have met them in my journeys in other years. My garden is covered. Only the tops of the tallest labels now mark the places of the roots planted in the heyday of July and of the seedlings that broke the ground in early May.

It is the first day of the buried hopeful year. Through the snow I have waded to my garden of dianthus. Reassuringly the staunch labels stand through the snow, marking the very places of those precious roots I have planted, and on some of them can be read parts of the names. Do the plants know the names are on the labels, that they may not forget themselves in the long deep sleep of winter?

Over the ground around the plants was placed an inch of soft old mold when the last blackbirds left us, and later the stalks from the Golden Bantam corn were laid lengthwise between the rows of pinks; and now the snow has drifted the spaces full and is on guard for me. Yet here and there a blue-green tuft peeps through,— Picotee and Grenadin and Pheasant Eye from England, perpetual carnations from Germany, and I know that just there are maiden pinks and marguerites from France, alpines from Switzerland, choice things from other European lands. Here are Chinese pinks that bloomed for me in the lost autumn from seeds sown in the open in May and that will flower again when the swallows come, more than two hundred kinds.

How lightly I name the countries whence they came as if they were only idle thoughts! And yet I see the waysides where I have found them growing in these lands, the church with the old bell and the pinks hard by, the stile in the wall, peasants in the walks, green fields and backgrounds of mountains and temples beyond the bounds of Christendom. And how lightly I name the kinds,— those pinks that have been evolved in some loom of mystery through a million years and whose destiny to change or to perish or perhaps to exist for a million other years is all unsuspected by me! How worldly quick I am with contacts of men and mountains in a thousand lands that I myself have never seen, and how rich with suggestions so faint that I know them not! Under the snow the pinks are still England and France, Switzerland and Germany, far China where I have collected them on wild hillsides, and many places that I do not know. But these pinks are also mine. Some of them will refuse me, for I shall not learn their needs; but by and large they are mine, rooted in my soil, alive to my fingers, and they will not withhold their bloom from me, the blooms that first developed in some far-off place where children speak a strange language, but the pinks speak all languages.

Often I shall go to them in the long quiet winter, and shall remember. I shall stand in the snow and know they are there, snug and secure where they allowed me to place them. And when the creeks begin to run wildly in the springs and the fox-sparrows pause in their passage, I shall remove the stalks, and behold, a magic of blue-green tufts sitting primly in their rows!

The long northern winter holds the memories of the summer months secretively under the snow, and when the snow goes off we feel that have met old friends of other years that have been on a long journey. It is not so in southern lands, where the winter holds no secrets and uncovers no mysteries. All plants alike survive if they are more than annuals, and there are no memories. But in northern lands there are outstanding differences. Of all the herbaceous humble plants, perhaps the pinks hold their character best through the winter days.

Other pinks are in their little pots sunk to the ground-level in the heatless glasshouse, small and tender plants that might have lost themselves in torrential rains and sleeting snows of the open ground. Old moss was sifted over them when the weather closed, they were well watered, and I left for the winter. Now they are frozen in. But today I have carried snow to them and covered them, both for protection and naturally to supply the moisture they still may need.

When the robins are ready, these and all other pinks will be dressed for the new growth and I shall watch for every bright and spicy flower. Yet I sometimes think I like the pinks best when they are not in bloom, but when growth is quick and joyous in them and each kind takes its due shape and presents its best dress of different leaf and sheen. In bloom anyone can see them, and even the crudest hastiest visitor will note the differences; but when they are not in flower, nobody comes to see and they are mine alone. I know them as they are. When the thaws come in January and February I shall see them, every one, and they will have somewhat to say to me. There will be the flat green soft-leaved carpet of Deltoides and Graniticus, tussocks of Alpestris and Arenarius, broad low close mats of Petræus and the coarser blue-green of Cæsius, the deep loose grassy-blue clumps of Plumarius that I remember from my mother's garden, coarse sprawl stools of Caryophyllus, sods of Gallicus, stiff grassy heads of Atrorubens, Croaticus and Cruentus, the big-leaved loose tufts of Barbatus, the ragged stands of Laciniatus and Heddewigii and others of their sophisticated tribe.

This is my evergreen garden, under the snow.

December

It is now the high December.
The last betokened ember
Of the striving vivid year
That survived the brown November
Lies dead and painless here
Lies dead and pinched and sere;
And the fruits of proud September
Are hanging hanging here
They are hanging thin and sere;
And the masks of ward and rober
That bedecked the dyed October
They have found their finish here
They are lying crisped and sere
They are drifting bleached and blear.

It was in a far December
As distinctly I remember
Of a youthful doubtful year
That I sat in whitened fear
Of the death-end of the year;
For in forests gray and sober
I had mourned the red October
I had grieved for forests dry and drear;

And the crows and chickadees
And the wind-gusts in the trees
Made my sorrow sharp and clear,
And the leaves keen-edged and sere
Rasped an anguish in my ear
Of the dead and absent year.

But oh! the winds of great December
Since the dumb days I remember
Have blown me wholeness of the year
They have brought their tokens here,
And the proudness of September
Lays its best expression here;
And the silence seemeth good,
And the bareness of the wood
Is the bareness of the truth,
And age and youth
Do pause awhile and rest
At the glory of the East and the honor of the West;
And the year is never wanting
And the way is never vain
And the creatures go undaunting
In the windrift and the rain.

Blow ye snows of old December
Drifting drifting down
Blow ye leaves of hale November
Drifting sere and brown,
All the years that I remember
With the snow come down.

VI

EPILOGUE

Marvels at Our Feet

I came at the end of a long journey to the llanos of the middle Orinoco, in Venezuela. My companion and I had seen ranges of mountains and alluring summits. We had seen rain forests choked with tropical verdure and punctuated with gaudy-colored birds. We had tramped in many other places where the vegetation was abundant and fascinating, for we were botanical collectors, both of us.

Today we were away from our burros. We had dismounted in the shade by a cool stream. It was a tributary of the mighty Orinoco River, and in comparison was tame and unexciting. Yet we found many good things along that shore.

I saw a tiny island in the stream, a bit of flat land that had been surrounded by high water. It had no attractive vegetation as one saw it from the shore, merely a low green cover of grass-like things.

When I waded out to this islet I found it was three feet wide and ten feet long. I got on my knees and began to gather small plants I had not found before. Some of them were only an inch high. Once apparently the area had been grazed, perhaps before the high water came. The plants had adapted themselves to this circumstance by blooming and seeding at these puny dimensions, and this was the first of the wonders.

One of the plants that I found on that small island was the only one of its kind I found in Venezuela. It was a pigmy thing, as thin as a thread and less than four inches high, so frail that you could not measure its worth in gold. I suppose that the reason why I found it here and not elsewhere was because my eyes were close to the ground and I was intent on every tiny living thing. We are inclined to look too far away for treasures. We stand too much aloof from what in our superiority we call trivial things.

Yet far away in that lonely bit of South America, where probably no collector ever went before or will ever go again, that small plant lived its own life successfully, made its seeds and trusted them to the kindly earth, and was blessed by sun and night and rain. That island will always be a green place in my memory for the things I found there and emotions I felt.

WE GIVE TOO small thought to the beauty and quality of life and its products. It is amusing to hear so many persons always saying how many or how much. Every town has something bigger or deeper or taller or more costly than any other town on earth. Yet the earth among its company of suns and planets is very small.

Lately, in a meadow not far from my home, I saw a weedy, "common" plant, a buttercup. The night's rain had beaten it to the earth. It was a tall buttercup, and now it was lifting after the storm. You have seen the same thing happen many times.

It requires only a glance to see that here is a problem in mechanics and in strength of materials to baffle all calculation. The power must reside within the plant itself, for it has no external braces, derricks or hoists. This particular buttercup was three and one-half feet tall, broadly branched, heavily laden with seeds. Three and one-half feet tall with a stalk that at the ground was only three-eighths of an inch in diameter. Its height was one hundred and twelve times its greatest diameter (and half this diameter is a hollow cylinder), yet this plant had already raised itself considerably from the wreckage of the storm.

You have perhaps gazed in amazement at the skyscrapers of Broadway. It seems impossible that a habitable structure could be reared so high, and equally astonishing is the flexibility of this soaring tower, whereby it "gives" before the wind.

Yet comparison with the buttercup plant is startling. The Woolworth Tower is about 70 feet square where it rises from the roof at the fifty-fourth floor; if the buttercup proportion were carried out, the building would stand 8,400 feet high rather than 792 feet, or more than a mile and a half high.

It would have projecting arms without brackets or braces, reaching beyond both rivers bounding Manhattan Island, and these arms would be capable of swinging 180 degrees or more in the wind.

Moreover, when prostrated by storm, the building would have power to pick itself up. Back and forth in the cylinders of trunk and main branches great freight elevators could move up and down, and in branches trains of cars would glide back and forth laden with folk and merchandise piercing the fire walls at the nodes.

Everyone should be put in contact with the mystery of life that stands stark before us but which we do not apprehend. It is in every leaf, every growing thing, every pulse of life, every foot of earth on the planet. The rapture of life grows as our knowledge grows. It is unnecessary to go to far-away places to see marvels. The mystery is always at our door.

LET ME TELL you a story. It was in a Taoist Temple far away in Honan. Priests performed strange rites before the idols. Every utensil, every rude piece of furniture, the food, the torturing heat and wet of that climate, the outline of the clustered buildings, the speech and the conduct of the people of that far borderland—all seemed outlandish to an intruder from the West.

Night came down like a pall of loneliness over the shaven hills, the doors were bolted, the morning was far away. Even the plants, the birds and the insects were strange.

But there on the old stone temple wall grew the catnip, the same that is under my window in America, the same that has greeted me in many wanderings in other lands. What memories it held; what sweeps of the earth's surface were in its leaves and odor! Farmyards and castles, fields at evening, walks where every soul was a stranger, walls and ruins, lost days of youth with stories by the fire in the evening, memories of all the years that have crowded each other so fast—these were all in that catnip plant that grew in the chink of an old wall of a temple in China.

Some people who happened to pass that way expressed surprise at the strange plants we had found. I told them there were strange plants everywhere, strange because nobody saw them. I told them how Thoreau had found all the plants of the Arctic in his own back yard.

Lawns everywhere are great places for humble and amazing weeds to find asylum. One can learn great lessons from these unnoticed vital threads of life. They adapt themselves so admirably; they become so thoroughly a part of the conditions in which circumstance has placed them. Every weed has a meaning in the universal plan.

The proper objects of study are the things one oftenest meets. The study of Greek is no more a means of proper education than the study of a foot of earth, teeming with life. Anything that appeals to a man's mind is capable of drawing out and training that mind. I think that in the good time coming the teaching of geography will not begin with a book at all. It may end with one. But it will begin with the earth, and the products of the earth, in the very neighborhood in which the child lives. And it will include, with a certain sympathy, the weeds.

IT IS OFTEN said that a weed is a plant out of place, but this is not so. Nothing is more in place than a weed! Wherever they are, they make themselves at home. They are good colonizers; they make the most of their opportunities; they are always on time, and they have other commendable qualities. They are great schoolmasters; they have kept man going from the time he first began to till the earth, or until he gave up, and then they came and took possession and kept the ground in condition until another man came.

The earth would be bare and bald indeed without them. Man tills only a small part of the land he has cleared. He leaves much of it furrowed, scarred and hard featured. The weeds of one kind or another take up the task of recovering it. The gullying stops. Mold forms. All the biological processes of soil building go forward. The waste places around man's slovenly habitations are covered. We speak of ragweed and plantain as ugly plants, but this is only because of their association with man's unkempt by-places and wastes. I would rather see a good pigweed than a poor rose bush.

Travellers never see the weeds, and we who farm or make gardens must kill them, yet they are messengers sent around the world, the foreguard of comradeship, the perfect adaptation to all the conditions and needs of life. Everywhere that I have been I have looked for the most common of them, so little noted that they are uncommon. The great views, the resplendent objects do not make a man at home in the world and content, not until the trivial and the common have a meaning to him.

Catnip was not the only old acquaintance I found back there in China, far from the thoroughfares of travel. In the yard of that temple were plantain, wild carrot, the sprawling mallow and the black night-shade. Nearby were smartweeds and docks, foxtail grass and pepper-grass; and in one corner was a lusty plant of fennel, the same fennel that I found growing shoulder high in my garden when I came home. I was not so far away, with so many good friends to meet me.

The weeds will never release their guardianship of the earth's surface. They will contend with men wherever men break land. As the contest goes on, they will fight every year the harder. I have no fear of the outcome. I like to think of farmers and gardeners, in the long conflicts, as conquering men.

WE MAY AS well, I think, accept it first as last that the farm is not the place from which to derive great money incomes. We should soon be ready to turn from our false gods and to capitalize the riches that may belong to a gardener, and to nobody else. The art by which from day to day one may grow, oneself, in country liv-ing; the appreciation of a separate life; the love of beautiful sur-roundings and of all those gracious solitudes that lift the inner mood—without these, there is really very little incentive to live in the country; without these, country life is tame.

It is not enough to be comfortable and make money. There is no satisfying project that does not run beyond the cash in hand. There is no relish in households not well managed, no joy in gardening in gardens not well kept, no gratification in work not well performed. And there is no pleasure in leisure that is barren of wonder and unproductive of those enthusiasms that warm the heart and burn in the brain.

Today a man told me that he had acquired a bit of land. A gar-dener, he said, had told him that the deeper you dig in the earth the nearer you get to Heaven. He had tried it. "It's wonderful, wonderful!" he exclaimed. I remember another gardener whom I asked, "How much land have you?" He answered with pride that he had one acre, and added: "It's a wonderful acre; it reaches to the center of the earth in one direction and it takes in the stars in the other." The same is true of any square foot of the earth.

To COMMON MINDS common things are not wonderful. Such minds have no desire of inquiry. They never grow. The well-trained mind probes beneath the surface, and wonders at everything; and this wonder, grown old and wise, is the spirit of science.

Go where you will—to lawn or roadside, to a clean-kept garden or cornfield, to a forest or to the torn but slowly mending slash of a railroad embankment—to anywhere where there is green, and you will see at your feet, endlessly continuing, a terrific and various battle.

Life is a fight in which most plants die young. So intense is the competition for every foot of the earth's surface that a man can seldom take a step on open ground without crushing or disturbing a whole cooperating plant society.

Plants must live together. Therefore, like people, they associate. They contend, but also, they become adapted or accustomed to each other. There is one association of plants for the hard-tramped dooryard—knotweed and broad-leaved plantain with interspersed grass and dandelions; one for the fence row—briars and choke-cherries and hiding weeds; one for the dry open field—wire grass and mullein and scattered docks; one for the slattern roadside—sweet clover, ragweed, burdock; one for the meadow swale—smartweed and pitchforks; one for the barnyard—rank pigweeds and sprawling barn grass; one for the dripping rock cliff—delicate bluebells and hanging ferns and grasses.

Every plant has good reasons for growing where it does, rather than elsewhere. Try to find those reasons, and if you get no answer, do not cease to wonder, for plants hold as many mysteries as the soil and the air and the sunlight from which they are formed.

In my youth we thought that we knew all about the soil, or as much as farmers need to know. Soil was then—according to the teaching of Liebig and the early soil explorers—a reservoir of chemicals. Next came "soil physics" and new ideas were placed before us. We began to talk about the soil particle, and structure and texture.

THEN CAME THE biology of it, with the discovery of micro-organic activity, and we visioned a living soil. Now we are coming to an understanding of colloids, but it is not to be supposed that we are yet in sight of the end. The next generation will have a very different conception of the supply of soil nitrogen and of fertility in general.

Thin like the skin of an apple, the soil layer of the earth has been formed through countless ages of weathering and the accumulation of organic remains. The history of the planet is recorded in it. Whatever may be at the center of the earth, we know that this thin exterior supports the life of the planet, and that this is the arena on which the drama of civilization is acted.

Among a throng of other plants, scattering back from raw earth at the cut of a railroad, I find one plant of a certain kind, pressing close to earth, and bearing fruit—a wild strawberry. The thin fragile tissue of its leaves is cut to pattern, with all those veins and margins and minute hair coverings that distinguish it.

We do not know why. We only know that all stalks and twigs yield leaves and seed, no two identical in all the world, but unerringly arranged to a recognizable pattern, after their kind. Seeds produced in prodigal number, each one perfect and ready, seek to reproduce. Most of them die, thwarted. Millions are lost sight of completely. They lie with refuse and grains of soil but they never forget their breed, and if they do grow they produce no other kind of plant whatsoever. This is one of the major mysteries.

This strawberry plant has triumphed. It has lived. It is bearing fruit. At first sight there is very little in that to suggest kinship with the vast transparencies of sunshine and air.

Yet this stem, this leaf, this fruit are all of sun, soil and air compounded. This small scarlet fruit is oxygen, it is hydrogen, it is carbon, the carbon you see in smoke. It is nitrogen, a gas that you have not perceived as such, although you live always within it. It is calcium, magnesium, phosphorus; it is iron, the same iron that is in a locomotive. It is potassium, and other elements besides.

THIS FRUIT IS water—water delicately and deliciously flavored with many intricate compounds. Perhaps it is nearly ninety per cent water, yet so nicely is the fluid held in fiber and cell that revolve it as I may, gently, and the water does not spill.

Inconceivable are the energies locked in the unyielding earth and the invisible air! Flowers and herbage all perfect and shapely and colorful, wrought by the countless addition of cell on cell, are whipped away, broken by storm, trampled by cattle, as if they were the very riffraff of creation.

Numberless millions of insects—matchless in shape, complete to the last detail, an amazing exactness and adjustment of circulatory, digestive, nervous and sexual systems—dance for a moment in the sun, then are gone forever. And no account is kept. The millions of men have come hopefully into being and have forthwith passed away; and nobody knows where they have gone. The legions rise, march their little day, and perish. Others take their place. The pageant never halts.

See now how these leaves of this small strawberry plant stand forth extended to bathe themselves in light. Delicate, invisible forces of life are moving mysteriously in these thin tissues, hidden away in laboratories themselves invisible, every laboratory and every process as perfect and fully furnished as if each leaf were to last forever.

These leaves will die. They will rot. They will disappear into the universal mold. The stuff that is in them will pass elsewhere, perhaps to the egg of a newt, to the root of a tree, to a fish in a pool. The energy will be released to reappear, the ions to act again, perhaps in the corn on the plain, perhaps in the body of a bird. The atoms and the ions remain or resurrect; the forms change and flux. We see the forms and we mourn the change. We think all is lost; yet nothing is lost. The harmony of life is never ending.

Every man knows in his heart that this is so. Every man knows in his heart that there is goodness and wholeness in the rain, in the wind, the soil, the sea, the glory of sunrise, and in the sustenance we derive from the earth.

We deceive ourselves if we turn from the essentials and try to satisfy ourselves with the small and trivial gratifications of this age. Let us look more closely about us and see how good are the common things, how marvelous are all things made at the beginning. The meaning of life is in its beauty. And ten thousand years from now children will call across the centuries that the world is young, that the sunshine is good, that love and faith, and mystery and the buoyancy of life are the only realities.

Society of the Holy Earth

I propose a Society of the Holy Earth.[1] Chapters and branches
it may have, but its purpose is not to be organization and its
practice is not to be the operation of parliamentary machinery. It
will have nothing to ask of anybody, not even of Congress. It will
not be based on profit-and-loss. It will have no schemes to float,
and no propaganda. It will have few officers and many leaders. It
will be controlled by a motive rather than by a constitution. The
associations will be fellowships of the spirit.

Its principle of union will be the love of the Earth, treasured
in the hearts of men and women. To every person who longs to
walk on the bare ground, who stops in a busy day for the song
of a bird, who hears the wind, who looks upward to the clouds,
who would protect the land from waste and devastation realizing
that we are transients and that multitudes must come after us, who
would love the materials and yet not be materialistic, who would
give of himself, who would escape self-centered, commercial and
physical valuations of life, who would exercise a keepership over
the planet,—to all these souls everywhere the call will come.

What may be the opportunity to express oneself in the public
interest, I do not know; but at least one may be ready, as the earth
is ready, and I would stimulate the desire. Here is the beginning of
universal service: the hope of humanity lies in universal service.

Appendix I
The Garden Fence

Horticulture, the art, is old. It had its origin, with twin agriculture, in the fertile valleys of Asia, while yet the world was new. Man early learned to till the soil. He was a farmer. The earth gave him her fruitage. He selected and improved it. Generation after generation the slow increment of progress accumulated. The fruits of the first garden gave place to others. Gradually the old were lost, and the best were scattered to the four quarters of the globe with the early migrations of men. The history of many of our cultivated plants is almost a history of the human race. But with the gift of fruits, God sent other friends, disguised. Weeds originated when cultivation originated. There are no weeds where there is no cultivation. They are enforcers of duty. They early punished neglect with the consuming growth of tares. They have always been coercers of improvement. It is singular that we do not recognize this fact. Even Virgil was alert to it: —

> The father of human kind himself ordains
> The husbandman should tread no path of flowers,
> But waken the earth with sleepless pains.
> So pricketh he these indolent hearts of ours,
> Lest his realms be in hopeless torpor held.

....

.... All these things he did
That man himself, by pondering, might divine
All mysteries, and in due time conceive
The varying arts whereby we have leave to live.

Surely ours is a goodly heritage. Until our time has man improved upon nature, till the first parents of cultivated plants are lost, and we are bewildered with endless variety. If we cannot discover the devious path by which every fruit has come through the centuries, gathering here and there an element of that mysterious something which better fits it for the use of man, we can, nevertheless, enjoy an heritage which surpasses the hanging wonders of Babylon or the fabled gardens of the Hesperides. Perhaps we are approaching the limits of this development. Certainly our methods of cultivating are not essentially different from those which find record in Columella or in the verse of Virgil, methods which in essence were old when those authors wrote. The ancient art appears to have taken on a fixedness which is indicative of staid old age. We plough and sow and reap as did our fathers. If we reap more than they, it is chiefly because we have improved a little more in the line of their improvement. Surely here is not a field for the impetuous Yankee, who would conquer countries of which his father had never heard, who is irrepressible in any enterprise which promises profit, and demands business, brass and brains.

In 1795 a short and unpretentious article on grafting appeared in the Philosophical Transactions of England. The writer had observed that in England the most disastrous of the diseases of the apple and pear was the canker,—a browning and dying of the younger shoots. It was the common opinion among orchardists that this disease is caused by a deterioration of the variety; the older varieties were running out. The writer opposed this view, and assumed that the disease had been conveyed, in each particular instance, by unhealthy scions. He conducted a series of experiments. He procured healthy young stocks, and grafted upon them the brightest and thriftiest scions he could secure from the cankered trees. When these had grown, he inserted the best scions which they afforded on other fresh seedling stocks. This progressive operation was repeated for six generations. Although he did not escape the canker, he found that he had hit upon a fertile trail. He satisfied himself that scions from old and worn-out trees are prematurely productive and short-lived, and reasoning from this he concluded that scions from very young seedlings would prove to be tardily productive and long-lived. Numerous experiments appeared to prove the proposition. Scions maintain their essential characters when set upon other stocks, or at least the characters of growth and fruitfulness. The graft will probably not endure long after the natural expiration of the tree from which the scion was taken. Probably most of the ancient varieties of apples had been propagated from scions from old and feeble trees, and, as a consequence, most of these fruits known to Parkinson and Evelyn had become extinct. The direct and impartial statements, the scientific methods, and the novelty of the subjects treated, at once brought the paper and its author into prominent notice.

Four years later our author appears again. Again he is a pioneer. The canker in apple and pear trees still demands his attention. He had observed that in the animal world in-breeding produces disastrous results. May there not be something akin to this in the vegetable kingdom? He proposed to cross-fertilize one variety of apple with another, hoping from the seeds of the cross to secure new and healthy varieties. Impatient for results in a field entirely new, he began experiments with pease also. The progeny of the crosses were new and peculiar, and the details of the experiments are still full of absorbing interest. At this time the whole manner and method, the whole physiology of the phenomena of pollination and fecundation were unknown. Numerous doubts arose in the mind of the experimenter. He endeavored to ascertain if one seed could be the product of two males, if the quantity of pollen used exerted a varying influence, if the male or the female parent is the most potent, if successive crosses would still change the offspring, if the characters originating from crossing can be discharged by subsequent culture. He experimented with apples, pease, wheat, grapes, and other plants. We who are familiar with the magnificent science which has to do with the crossing of plants, which first took definite shape and direction under the genius of Darwin, and which in its phenomena and influence is boundless, are fully prepared to admire those men who first caught a glimpse of this wonderful plan of nature. We look with a species of reverence upon Conrad Sprengel who in 1787 began to study in the fields the mutual relations of flowers and insects, and who became impressed with the idea that all parts of the flower subserve some definite economy; that "the wise Author of nature would not have created even a hair in vain." But in this same year 1787 a greater man than Sprengel began his work upon the same subject. The German and the Englishman, unknown to each other, caught the thread of nature's purpose, and began to unravel her close-woven fabric. The one interrogated nature in the field, the other courted her in the garden. Our author, the Englishman, was impatient to apply to the common uses of life the discoveries he had made, although he recognized, perhaps as fully as the other, their importance to the preservation of species in wild nature. He saw, too, the relation of the insect to the flower. "Nature seems to have wished that no flower should be fertilized by its own pollen," said Sprengel,—a statement which has become celebrated. "Nature intended that a sexual intercourse should take place between neighboring plants of the same species," said our author,—a statement truer than the other. "Nature abhors perpetual self-fertilization," said Darwin. Our experimenter gives a pleasant account of the agencies of insects in cross-fertilizing plants. But after all he saw more clearly the relations of the phenomena of crossing to the much-loved plants of his garden, and ventured the assertion that "by this process it is evident that any number of new varieties may be obtained; and it is highly probable that many of these will be found better calculated to correct the defects of different soils and situations than any we have at present; for I imagine that all we now possess have in a great measure been the produce of accident; and it will rarely happen, in this or any other case, that accident has done all that art will be found able to accomplish."

Among the flowers of his garden, our author became convinced that all the parts of the flower,—the showy petals, the stamens and the pistils,—are but modified leaves. Although he was not the first to conceive these ideas, he nevertheless arrived at his conclusions independently, for the studies of Wolff and the poet Goethe were then unknown in England. Upon this apparently singular assumption rests much of the important investigation of to-day. He studied the motion of sap in trees, and made numerous experiments, some of which proved that the ascent of sap does not take place between the bark and the wood, but through the wood itself. In 1811 he gave to the world the now familiar method of root-grafting, with which he had experimented upon the pear, apple, plum and peach. A year later he published a minute and interesting account of the movements of tendrils, a subject now made classic by the work of Darwin. About the same time he introduced a peach which he produced from an almond. In the same scientific and quiet spirit he discussed the causes which influence the direction of roots, the nature and extent of expansion and contraction in the trunks of trees occasioned by heat and cold, the parts of trees first impaired by old age, and a long line of vital subjects, always with well-directed experiments. In many cases he came near anticipating some of the beautiful generalizations which we now know as Darwinian.

But what is the significance of this work, and who is its author? Horticulture has became a science, and Thomas Andrew Knight is its founder! Science has climbed the garden fence. It is not enough that we plough and sow and reap as did our fathers. Unto us are given countries to conquer of which they had never heard. Here is work for the impetuous Yankee; work which is as boundless as time and energy. Horticulture, the art, is old; horticulture, the science, is new. To get our science from the field and the laboratory into the garden, is the problem of the age. We must demand it there. Therefore, I propose to speak to you about the Garden Fence, or what we don't know about horticulture.

The fence which stands between theory and practice is relative. It exists and it does not exist. It depends upon the position of the observer, or rather upon his definition of the word practice. This word practice is much abused. To one all knowledge is practical: it is a part of a grand scheme of progression, and, sooner or later, it exerts an influence upon some one or more of the varied industries which support the life of man. This is a philanthropic view of learning. It recognizes the important fact that all knowledge is practical, because it adds to the weal of mankind. Money is not always the true measure of the practical, else what is practical to the recipient is impractical to the giver. A person looked through a scientist's microscope; he saw the peculiar objects which were explained as the parts of a fungus, but he saw no application of the knowledge he had gained. "What's it good for; what's the use of all this study?" he asked with disgust. "It gets me a living, sir," retorted the scientist. We must not measure knowledge by its immediate effects, any more than we should measure an apple tree by the young seedling. But he who invariably measures the influence of education and knowledge by money, is a niggard, and is opposed to advancement. It is time we did something for the fun of it. If we are to make science conducive to the needs of man, we must search all science, for we know not where some treasure is hidden. The horticulturist will quite as often find some useful hint in an inconspicuous weed by the roadside as in the cultivated products of the garden. It was by experimenting with a frog that Galvani discovered galvanism. That frog lives in every industry which brightens our civilization. It was a wild geranium which gave Sprengel the hint of that wonderful kinship which exists between the insect and the flower; and that wild plant of the fields will always linger in the traditions of our science and our horticulture.

One can never become a successful investigator in any subject if his whole skill and education are confined to that subject. Much of our experimenting is entirely worthless, because the experimenter is not able to grasp the relations which exist between his subject and other subjects akin to it. And herein lies the greatest gulf between theory and practice. Says an experimenter, Prof. W. R. Lazenby: "Nothing that is really good or true in theory can ever fail in practice. If failure occurs, it proves that the theory is false or the practice incomplete." It is singular how loudly many men decry the opinions of scientists as vague and impractical theories, while they themselves are bristling with whims and notions that would do justice to the absurdities of the Middle Ages. If a thousand devils can dance on the point of a needle, how many stalks of chess will grow from one grain of wheat?

But after all there is a conspicuous fence about the garden. The botanist searches for plants in woods and glades and fields; he studies them; he chases them to the garden fence and stops! When a raspberry gets into the garden it is without the pale of the true science of botany! "Our roses have ceased to be a botanical study," said a great botanist, when in fact they have never been worthy so close a study as now, when they have run into all the forms of our gardens; when they have so far disguised themselves as to make their very origins matters of speculation. Why have they varied, how have they varied, how much can they vary, what is their relation to soil, to light, to heat, to moisture, to pollination from other varieties or species; in short, what does botany tell us of the rose under cultivation? Nothing. We don't know the meaning of a rose; if we did, who knows but that we should find a key to many of the secrets of the vegetable world? The botanist throws it aside because it has lost its permanent specific characters; he cannot name and classify the perplexing multitude of forms. But the very fact that the plant is so perplexingly variable is all the more reason why the botanist should aid us in its study. Said Darwin: "One new variety raised by man will be a more important and interesting subject for study, than one more species added to the infinitude of already recorded species."

We must get below the surface indications. We need to know the principles which underlie our experiments before we experiment, or else we must experiment for the purpose of discovering the principles. Experiment is rife to-day; the empirical spirit of the age is contagious. Every one experiments or investigates. The greater part of this experiment is the reflex—the echo—from the scientific tendency of the times. It commonly has little scientific basis and no permanent value. People are experimenting to find out what they ought to know without experimenting. Every experimenter must know what experimenting has been done already. He must be an educated man. Experiments are often interpreted incorrectly; they are said to teach what they do not teach. A person sows land-plaster on one-half his wheatfield and leaves the other half unfertilized. Upon the plastered portion the wheat is four or five inches higher than on the other. Therefore, says Quizicus, plaster produces a great increase of wheat; not thinking, however, that growth of straw is one thing and yield of wheat another. A gardener had two rows of onions. Upon one he applied guano, upon the other bonedust. One yielded four bushels more than the other, and he attributed the larger yield to the fertilizer; but under the same treatment they would, undoubtedly, have varied as much. An observing fruit-grower possessed a plant of smooth-fruited gooseberries. A favorite family cat, having unceremoniously died, was buried underneath the bush, and behold! the next year the bush bore hairy berries, and has so continued to do until the present day! Most of my neighbors keep seed corn by stripping the husks and braiding them together and then hanging the ears in a dry loft; but one, more acute than the rest, one year hung his corn in a hoghouse, by way of experiment. The next year his corn failed to grow; therefore, said he, corn hung in a hoghouse will not grow. This is akin to the valuable experience of a certain Irishman, to whom rhubarb was given in a case of sickness. He recovered. Shortly after, his neighbor, a Dutchman, fell sick, and Pat administered the rhubarb. The man died. Pat hastened home to write on the fly-leaf of his Bible: "Medicine which will cure an Irishman will kill a Dutchman." Surely, experiment is in the wind. Even the city editor has caught the contagion and writes: "I am building up an article on potato rot. What insect causes it? How does the rot get in its work? Is it more prevalent when cholera is raging?"

Of a surety, we need our botany and chemistry and zoology and meteorology in the garden. We need intelligent investigation. Moreover, we need extensive and extended investigation. If we need one thing more than another, it is that the botanist shall climb the garden fence and include within the realm of his science all the plants which we till. Even Knight made this demand, nearly a century ago:

> I cannot dismiss this subject without expressing my regret that those who have made the science of botany their study, should have considered the improvement of those vegetables which, in their cultivated state, afford the largest portion of subsistence to mankind and other animals as little connected with the object of their pursuit. Hence it has happened that, while much attention has been paid to the improvement of every species of useful animal, the most valuable esculent plants have been almost wholly neglected. But when the extent of the benefit which would arise to the agriculture of the country, from the possession of varieties of plants which, with the same extent of soil and labor, would afford even a small increase of produce, is considered, this subject appears of no inconsiderable importance. The improvement of animals is attended with much expense, and the improved kinds necessarily extend themselves slowly; but a single bushel of improved wheat or peas may, in ten years, be made to afford seed enough to supply the whole island, and a single apple or other fruit tree may, within the same time, be extended to every garden in it.

There are a few who have surmounted this garden fence at some of its highest points, and of these, none stand out so clearly as Charles Darwin, the grandest horticulturist of any generation,— the man whose work pervades all scientific thought to-day. It is not the man who tills the soil who is necessarily the best horticulturist; it is, rather, he who knows nature best, and who can put his knowledge into form for others to use. A Darwin, although he never held a hoe, can do more for the permanent and profitable advancement of horticulture than all the horticulturists of New England. Out of this great wave of unscientific experiment which floods our land, we shall one day expect another Darwin to rise, who shall reveal to us more of the methods of nature than we can dream of to-day.

The art, the handicraft, of horticulture is well understood; but every part of it which touches a science demands further investigation. We do not know the scientific principles which underlie these handicrafts.

Of the subjects of science which have been worked out, I know of none so thoroughly done as pear-blight. Indeed, the researches of Burrill and Arthur, during the last five years, may be taken as the type of successful investigation regarding the diseases of plants. We hear much, nowadays, about parasitic fungi and their action upon the plants of our garden, and in many cases we can apply efficient remedies or preventives. We are inclined to regard the whole subject as one well understood, while, in fact, very few are so imperfectly understood. We have not yet been able to describe, to become acquainted with, the outward appearances of many of these fungi, and in comparatively few cases do we know the whole intricate round of life of the species. But we must soon begin to learn another set of facts; we must discover the relations which exist between the nature of the lost plant and the aggressive fungus. Why is it that the red-rust always attacks the Kittatinny blackberry, while some others sorts are exempt? Why does the bean-pod fungus attack the white wax variety in preference to others? We say that one variety has a thicker epidermis than another; that it is a more vigorous grower, and is therefore enabled to resist the attacks of the fungus; but these notions are indefinite. The fact is, we don't know why one variety resists a fungus and another does not. If we did, one of the problems of our horticulture at present would be the breeding up of fungus-proof plants upon scientific principles. If there is any attempt in this direction at present, it is entirely haphazard. Not many years since, the notion was entertained by many scientific men that the peculiar objects which we know as parasitic fungi were not distinct organisms, but simply modified cells of the diseased plant. We have now outgrown this notion; but we are, nevertheless, far short of solving the mysterious relations which exist between the fungus and the plant upon which it grows. We have emerged from one difficulty but to encounter another. The things which we do not know about horticultural science are astounding in number and importance, and they pertain to the commonest operations of the garden as well as to the most difficult and extraordinary. Let us examine, for instance, the simple matter of grafting and budding, which, so far as the art is concerned, is as well understood as tillage itself. It was practiced by the Romans. We bud our fruits as they did, but we know little more than they concerning the principles of the operations. What do we know of the laws of affinity between plants,—laws which enable us to determine the limits of grafting? Some pears thrive upon the quince, some do not; but the quince does not thrive upon the pear. The pear is short-lived and unsatisfactory when grafted upon the apple, which is very near it in botanical kinship; but it does just as well, it is said, upon the thorn, which represents a distinct genus. The peach takes poorly on the apricot, but it and the apricot thrive on the almond and the plum. Most plums do well upon peach roots, but the Canada Egg commonly fails to unite, and the Lombard makes such an imperfect union that it soon breaks off; still, between these plums and others, we can discover no differences to account for these peculiar behaviors. Sweet cherries do well on the Mahaleb cherry, but the Mahaleb will not succeed on the sweet cherries. The gooseberry will not grow on the edible currants, but it thrives well on the wild buffalo currant of our West. A certain Chinese orange almost fails to bear upon its own roots, although it becomes very prolific when grafted on one of the lemons, a distinct genus. We know scarcely anything of the influence of stock upon graft, and we are unable to discriminate, in most of the recorded facts, concerning the matter, as to whether some change in the scion is produced by the stock upon which it grows, or by soil, climate or culture. Still, the subject is one of immense practical importance. We may have a tree with plum roots and almond leaves, and the trunk may be composed of both peach and apricot, but we have no knowledge of the physiological relations which exist between the parts of this composite individual. We have a few facts concerning some indefinite influence which the scion exerts upon the stock. An experienced nurseryman habitually looks ahead, when he is digging trees, to note the character of the tops of the trees he is about to dig, knowing that a very upright grower will have a tap root, and a very bushy grower a spreading root. But the top is the scion and the root is the stock: how is it possible that the scion can influence the root upon which it grows? Many shrewd nurserymen tell us that if we graft a plum upon the young root of a peach, in a few years the peach root will change to a plum root, the identical fibers which were once peach become essentially plum in their external features. Variegation has long been noticed to be an occasional influence of scion on stock. A stock with ordinary green leaves is sometimes forced to produce variegated leaves by inserting a bud from a variegated variety; and this is all the more singular from the fact that the buds themselves often fail to grow. The stock may be influenced in this wonderful manner below the insertion of the bud as well as above it. If we attempt to explain this mystery we but unlock other mysteries. What is variegation? What causes it? Some contend that it is a contagious disease, and that budding is an inoculation. I find that so long ago as 1727, this idea was advanced; for Prof. Bradley of the University of Cambridge in England observed,

> that the distemper which shows itself in the yellow and white variegations of the leaves of the common white jessamine, and several other plants, may be communicated to every plant of the same tribe, by inoculating only a single bud of the variegated kind into the others which have plain green leaves; and, though the bud does not live, yet barely by the application of it to the healthful tree, we shall find the yellow blotches or variegations of the unhealthful bud communicated to every part of the healthful plant. Just as it happens when a man has had the small-pox inoculated upon him, his whole mass of blood will become infected with the poison.

We are little wiser upon this point than Bradley was. Now there is on record a case in which an entirely distinct plant, once regarded as a true natural species, was produced by grafting a scion of one species upon a stock of another. A hybrid was produced by grafting. Did we know how and why this came about, might we not apply the principle indefinitely? Now we have the remarkable statement that a certain Italian, through long study and experience, has hit upon a device by which he can produce new varieties of roses by the simple art of budding. Whether or not this statement be true, it is, nevertheless, a straw which indicates a current. The more we study this apparently simple matter of budding and grafting, the deeper we are surrounded by an impenetrable maze of mystery; we are everywhere limited by the unknown,—unknown. I am not to be understood as saying that we have made no advancement in grafting. Columella declared that he could grow several kinds of grapes in a single cluster, by tying together cuttings from four or five varieties, enclosing them tightly in an earthen tube, and burying them in the soil to grow together. In this manner he said that a compound vine could be produced, which would bear many-fruited clusters. He would produce seedless grapes by splitting his cuttings, removing the pith, and then placing them together again! He also detailed a device by which "scions of all kinds may be grafted upon all sorts of trees whatsoever." The Romans evidently had little notion of the affinity of species. Virgil would produce a curious medley:—

> But thou shalt lend
> Grafts of rude arbute unto the walnut tree,
> Shalt bid the unfruitful plane sound apples bear,
> Chestnuts the beech, the ash blow white with the pear,
> And under the elm the sow on acorns fare.

We should expect that the horticulturists of to-day should not hold such notions as these; verily, there has been advancement, but for the most part it has been a stumbling advancement. Our Pegasus is blind.

Let us return to the botanist. Our curiosity is excited as we see him strolling critically over the fields, collecting-case in one hand, botany in the other. How does his botany help him in his rambles? Is it possible that he can identify all the multitude of forms of vegetation with names and descriptions? He can; and herein lies one of the wonders of botanical science. The classification and the method of naming are such that the diligent botanist can hold in his mind the names and the kinships of thousands of plants with no tax upon the memory. There is no system of arrangement so complete, no logical method of subordinating a lesser character to a greater so thorough, as the systems of classification and nomenclature which we apply to wild plants and animals. On the other hand, there is no system more bungling, none more thoroughly haphazard, than that which we apply to the plants of the garden.

Is there not some way to get our classification and nomenclature over the garden fence? If the subject is beset with difficulty, so much the more do we need system, and so much greater will be his honor who constructs it. I hope to see the day when the gardener can botanize intelligently in his garden; I hope to see a handbook which will aid us in the determination of garden varieties. And this, you must admit, would be an exceedingly "practical" sort of a volume. It would endeavor to give us the synonymy of each variety; it would tell us, before we make our spring order for seeds, whether the Leyden White Summer, the Satisfaction Black-seeded, the Black-seeded Yellow, the Fine Imperial Cabbage, and the Berlin lettuces, are in fact distinct varieties or whether they are all names for one and the same thing. This is an exceedingly important matter, to find out if many of our common varieties are really distinct, and to hunt out the oldest name for a permanent appellation. It must be investigated with great care, and upon a scale not profitable for the individual gardener, who must live by the sweat of his brow. It must be investigated by persons who have trained eyes. In this direction the New York Agricultural Experiment Station is a pioneer in this country, so far as I know. Each year the Station garden grows some one vegetable in all its varieties, for experimental purposes. In the Station report for 1883, 58 varieties of beans are accurately described and compared. Progressive horticulture demands that some efficient system of classification be worked out for each of our orchard and garden plants. It is by no means a satisfaction, if we wish to find the name of some apple new to us, to be obliged to know the name before we can find the name in an alphabetical arrangement.

We must learn the possibilities of native wild plants. It is in this direction that we must look, in many cases, for increased hardiness and productiveness. Our Wild Goose and Miner plums indicate a new field for advancement in plum culture. Our wild black currant and dwarf sand cherry are awaiting investigation. We must breed the bitter lining out of the pecan and the big seeds from the papaw. We often account it a fortunate circumstance that the cradle of the human race was rocked in southwestern Asia, the home of fruits, the land which flowed with milk and honey. But if the Garden of Eden had been in America, our heritage would have been as great, perhaps greater. The possibilities of our wild fruits as a whole are great. Already our gardens are planted with native grapes, native strawberries, native raspberries, native blackberries and native cranberries. The native species are by no means all utilized. A fertile field of future experiment will be the growing of edible fungi. Many wild species are agreeable and wholesome, but so far we have succeeded in cultivating but one, the world over.

Gardeners are familiar with "sports," those occasional mysterious plant forms whose advents are unknown until they suddenly appear. The phenomenon itself of sporting is known of late, since the work of Darwin, as bud variation, a term of great importance to the gardener, as malaria is to the doctor, since it covers volumes of ignorance. Cherry trees which habitually bear red fruit sometimes produce a branch which bears white fruit. Yellow plums have been seen on certain branches of a purple-fruited tree. Greening trees sometimes produce russet apples, and russet trees sometimes produce greenings. Potatoes are sometimes half white and half purple, and planting one side or other of the tuber will often reproduce the peculiarity of that side. Weeping branches appear on trees of upright growth. Variegated or curiously cut leaves appear suddenly on many plants. Plants so unlike all others as to be called distinct species have originated by bud variation. In this manner the moss-rose probably originated, and certainly the nectarine is a sport from the peach. We know nothing of the causes of bud variation. We shall expect to someday discover many and diverse causes for these fitful phenomena. Did we know these causes now, we might apply them to the production of better fruits. Sport is certainly a relative term. It is a sport to-day, because we do not understand it; to some horticulturist of the future it will be but the operation of a law.

We sow with the confidence that like produces like, that as we sow so shall we reap; but the keen observer sees in the offspring of almost any seed, when sown in considerable quantity, a wide variation. Indeed, no two individuals are alike, although they spring from seeds grown in the same fruit. Plants have individual characters just as clearly pronounced as do people, and so imperceptibly do these characters widen in all directions that we cannot say when any character ceases to be individual and becomes varietal; that is, common to a number of individuals; or even when it becomes specific or permanently common to a class. Thus it happens that characters which are in the judgment of one man varietal are in the judgment of another specific, or may be even individual. "Species are judgments," said a great botanist; and, necessarily, he who has the best judgment and the most experience is the best judge of character in plants. Such judgment is of supreme importance if one would enter the higher fields of modern horticultural research. Often the seeds from the same pod will produce plants very different in their characters; the seeds "break," as the gardener phrases it, and we get what we call new varieties. Why? We say that it is due to peculiarities of soil, of culture, of climate, of some previous influence of pollen, or something. Surely it is due to something; so far we are correct. Reasoning from this known tendency of plants to vary, people often construct curious notions, which lie entirely without the limits of possibility. These limits are readily distinguished by the botanist, but cannot always be detected by others. Here we find an explanation of those antagonistic notions which have been a feud between the farmer and the botanist; the notion on one side that wheat turns into chess, and on the other side that the supposition is absurd. It is curious to what extent this ideal transmutation of species is often carried. As early as 1747 a Latin dissertation, written under the direction of the learned Linnaeus, was published in Europe, to disprove the fallacy that wheat turns to chess. The notion has even an older history than this. The idea that certain grasses regularly transform into each other is as old as recorded history. It is said that in early times the peasantry of Europe had discovered a regular series of transformations, due to poor soil, from wheat to rye, then to barley, then to darnel grass, then to chess, and finally to oats! And it was also declared that the reverse conditions of a fertile soil would evolve wheat from oats through the same intermediate plants! At the present day, and in portions of our own country, chess and clover degenerate into timothy, and horse-hairs grow into snakes! And the people who observe these unorthodox pranks of nature are often among the first to scorn the idea of evolution, which attempts to account for the instability of species in a scientific manner.

We secure new varieties of plants largely by random. This method is unscientific, and, to the student of natural science, is unattractive. We do not know the possibilities which lie in a seed. Sometimes seeds contain two embryos, two initial plantlets. It was once observed that two per cent. of a lot of young Osage orange seedlings were united twins: the seeds had contained two embryos or germs, and the young plantlets had grafted themselves together. A still more remarkable case is that in which two very dissimilar plants were obtained from one seed of a fuchsia, the double-embryo seed, in this case, being the product of cross-pollination. We do not understand the mysterious effects of soils upon young seedlings. Prof. Tracy, of Michigan, sowed peas of one variety in a row which extended from poor soil to rich soil. Upon the rich soil he obtained a new variety of pea which reproduced itself from seed. We say that strong soil was the cause; but the same thing would probably not occur again in many years, under conditions which, so far as we could judge, are exactly similar. We do not know why some varieties or species of plants are more variable than others. Some cultured varieties will reproduce themselves with remarkable permanency from seeds, others will not. These fixed varieties, those which come "true to seed," we designate as something more than varieties; they are races. We have a name for them, as indeed we do for most of the phenomena of nature; and I often think that there is a tendency to crawl under these technical names, and to applaud ourselves with the idea that we have picked the meat out of nature's puzzle. Seeds from the young plants appear to produce a better and more variable offspring than those from old plants of the same species. Dr. Van Mons of Belgium, inspired by this fact, built for himself a permanent name in the science of horticulture. He selected seeds from the first fruits of young trees, especially from young trees of new varieties, and planted them. From the first desirable fruits of the seedlings obtained, he again selected seeds and planted, and so continued to do for several generations. Each succeeding generation fruited sooner than the preceding ones and produced better fruit, until about the fifth generation, beyond which there was no increase. The fifth generation of pears bore at three years from the seed. Van Mons proved that by selecting seeds from these young plants, which are in "a state of variation," whose characters are not yet fixed by age, we shall rear the best seedlings. And here another question arises. If the characters of young trees are not yet fixed, will the first fruits be the same as those which the tree will bear in mature years? Are the habits of the boy the same as those of the man into which he grows? We know that in many cases they are not. But here we find a fact that we should not expect from the conclusions of Van Mons; the first fruits of the tree, if they vary at all, are commonly inferior to the later fruits. Still these same inferior fruits give a superior progeny! Would it be possible by root-grafting scions from a seedling at different times during the first four or five years of its existence to secure different varieties of fruit? We shall try it.

Verily, we do not know the possibilities of a seed. We need well directed, extended experiment. We need to plant very many seeds of every useful plant, under conditions as nearly alike and as much unlike as possible, and to make a numerical record of the peculiarities of variation. Do they vary most constantly in this direction or that? We may then be able to discover some law of variation.

In a general way, we have hints as to some causes of variation; but here, as elsewhere, we are obliged to cover our ignorance by a technical term. Certain conditions of vegetation attend certain climates, and we habitually refer those conditions to climate as a cause. This disposition by no means discloses a specific cause, however. Climate is ambiguous. In common usage, it includes latitude, heat, moisture, drought, winds, intensity of sun's rays, electrical conditions of the atmosphere, and other phenomena. We must analyze climate, and study the effects of its component parts. Here is a field which is wonderfully fascinating, from the fact that it deals with problems of such magnificent proportions; it includes at once, within its scope, the whole world, with all its depressions and elevations, its currents and counter-currents, its land and its waters, its winds and its calms. It traverses every unknown country, under the lead of versatile Von Humboldt, the father of botanical geography; it visits the islands of the sea and climbs the awful ranges of the Andes and the Himalayas. On the other hand, it recognizes every local distribution of heat and cold. We are becoming familiar with some of the results of a change in latitude and climate, but we can scarcely frame laws. When a vegetable is taken North it usually becomes dwarfed. The average height of Indian corn in the Gulf states is twelve feet; in Canada, six feet. Compare our Yankee corns with the Southern dent. Many woody perennials of the South become herbaceous annuals at the North; castor beans and cayenne peppers are examples. The apples of northern Russia grow on bushes, rather than trees, which are planted in hills after the manner of corn. Aside from dwarfing, plants usually take on different forms as they are taken northward. The tops are lower and rounder. In lower latitudes, they incline toward a pyramidal or fastigiate shape. The lower branches of conifers are proportionally longer in Canada than in Carolina. There is evidently a greater tendency at the North for plants to sucker and to produce underground stems. Although northern latitudes induce dwarfing, the amount of leaf surface is proportionally larger than southward. As checking growth induces fruitfulness, we can readily understand that plants are commonly more productive northward, so long as the climate does not interfere with the health and maturity of the plant. As a rule, however, it appears that the fruit of any species increases in size as we go south, but the number of fruits to a given extent of plant surface is greater northward. A recent census gave the average yield of wheat per acre as 14.2 bushels in the upper ten Atlantic states, and 9.8 in the Gulf States, and 30.66 bushels of oats against 14.2 bushels. The latitude of the greatest productiveness of any plant is usually north of the latitude of greatest growth; *e. g.*, if a plant reached its greatest size at 40°, its greatest productiveness might be at 45° or 50°. If dwarfing produces fruitfulness, without producing serious concomitant evils, it is desirable; for while we may lessen the actual amount of production on each plant, we can grow many more plants to the acre. The increase in plants can be much greater than the decrease in individual production, but there must be a limit to profitable dwarfing. The most productive ratio of size of plant to the amount of fruit it bears, is an important and entirely unsolved problem. It has been stated that in England the most profitable ratio for wheat is about ten parts of straw, by weight, to seven parts of grain. Given, the profitable ratio and that latitude where this ratio will be naturally developed, and we have the essentials of a great advance in intensive horticulture. Seeds could be distributed from the given station; and even if we were not able to produce distinct varieties, which should possess this ratio as a permanent character, the seeds could be frequently distributed. We are gradually approaching this climax. Northern grown seeds are now in great demand. This fresh stock, this change of seed, is of great importance in many respects, of which the feature I have detailed is perhaps the most important, though the least understood and most neglected. By selecting seeds from a certain locality we are enabled, with a great degree of accuracy, to secure the salient features of the plant in that locality. "The enhancing of any peculiar feature of growth may be done by bringing seed from a climate which has that tendency."[1] Latitude, or some of the conditions of climate which accompany latitude, has a potent influence upon color. Northern fruits, like northern maidens, have ruddy cheeks. In old Russian song is a marvelous maiden, whose neck was like a swan, whose lips were like cherries, and whose cheeks were as red as the Volga apples. The object and the figure are attractive. The beauty of Alpine flowers is proverbial. On the unfrequented slopes of high mountains, fringing the perpetual snows, are the prettiest flowers the world affords. In vain do we search for the cause. It is pleasant to entertain the proposition of Wallace, that these bright Alpine colors are usually gaudy advertisements to insects, which are rare upon high mountains. The reciprocal relations of flowers and insects are always absorbing; but although the fact that Alpine flowers produce unusual quantities of nectar appears to uphold Wallace's hypothesis, we must nevertheless forego the pleasure of its entertainment. We find the same gaudy colors where insects are common; moreover, we can produce them, in short periods, by a transfer of culture. Perhaps we are beginning to solve the problem in the recent studies of the intensity of sunlight at high altitudes and latitudes. As we learn more upon this subject, we shall undoubtedly be able to control to a great extent the colors of our flowers. Indeed, Flahault had fourteen species of ornamental plants sown the same year in Paris and in Sweden, and of these, thirteen produced much brighter flowers in Sweden. The study of the intensity of sunlight will probably enlighten us upon the causes of high flavor in northern products, for be it known that high latitudes increase flavor in fruits; I am not able to verify here my above comparison in regard to maidens. We must know why it is that our apples and vegetables and corn are better at the North. It is lately asserted that even the watermelon, when well grown and thoroughly matured, is probably better at the North than at the South. If the world will still persist in accusing Brother Jonathan of trickery, it must, nevertheless, give him credit for honest, concentrated fruits. Hot climates develop poisons and aromas. Aromatic plants are characteristic of deserts the world over, says Wallace. I have in mind a pleasant incident of opening a bundle of dried plants which were picked twenty-five years before, in the deserts about Palestine, and so strong was their fragrance that it filled the room with "Sabaen odors from the balmy fields of Araby the blest." The historic hemlock of which Socrates drank, loses its virulence when grown in Scotland, and our sassafras loses its odor when grown in the cool summers of England.

Our studies of the relations of plants to climate must deal with acclimation,—a subject held in such different estimation by different observers, that while the eminent Prof. Lindley has great hopes for its future, Peter Henderson declares that "a life-time spent in the practical study of horticulture has forced me to the conclusion that there is no such thing as acclimation of plants." Corn, for instance, does not succeed in England. This diversity of opinion may arise, in part, from different understandings of what acclimation is. To one, acclimation means an entire change, a revolution in the constitution of a plant, so that it can exist in opposite extremes of climate; to another, it means a series of minor changes, taking place gradually, so that the plant can be cultivated or become naturalized through small but constantly widening circles of differences. The first notion supposes no limits to acclimation. So far as I know, it is unreal. The second notion, that of gradual acclimation and naturalization, is abundantly illustrated in every garden and by every roadside. It accepts the common observation that there are limits to acclimation. We cannot grow water lilies on a sand-hill or corn in a damp and cloudy climate. We are not able to say whether we can induce some entirely new change or series of changes to take place, in order that the plant may become accustomed to some radical difference in climate, or whether we simply intensify or draw out some latent tendency to variation, which exists in the plant in wild nature. Of the fact of acclimation, however, there can be no doubt. Plants adapt themselves to colder climates. In fact, the dwarfing consequent upon transference to higher latitudes is itself an adaptation, from the fact that the plant requires a shorter season in which to mature. The individual character of the plant is, in some instances, mysteriously changed. It is stated upon good authority, that twenty degrees below zero in Michigan is no more injurious to a given variety of peach tree, than zero in

Mississippi. We have numberless prophetic facts concerning acclimation, but of its possibilities we know almost nothing. Our science must climb the garden fence to solve the problem. It is possible that we must begin with the seed itself if we would acclimate. It has been thought that the reason why northern-grown seeds germinate quicker than others, in spring, is because the cold of winter produces in the seed an increased sensitiveness to heat and cold. Indeed, individuals of the same species were once kept, some in an ice-house, others in a warm cellar, and the former vegetated sooner and grew faster in spring than the latter. Upon this suggestion I am now experimenting with seeds, cuttings, and scions.

If we would fully understand the laws of variation of cultivated plants,—whether the variation is in the direction of acclimation or otherwise,—we must know the origin of the plants; we must know how they have varied in all previous times. The origins of many of our cultivated plants are lost in the mists of antiquity. They antedate civilization; they sprang from untaught nature, coincident with man. The primeval ancestors are lost. We search the records of every ancient people, and our perplexity is often rather increased than diminished. Sometimes history is altogether silent. How, then, can we know the unrecorded past? If man, by cultivation, has evolved our plants from wild nature, why cannot man, by a reversal of that cultivation, breed back to the originals? The common radish is unknown in a wild state. When radishes become spontaneous, or self-sown, about the borders of the garden, they lose many of their valuable characters. Their roots become somewhat smaller, much tougher, and the aspect of the plant is changed. Three acute observers—botanists necessarily—observed that the variations of these self-sown plants are in the direction of a certain so-called wild radish, which is a weed in poor soil, along the Mediterranean and in some places on our own Atlantic seaboard. This plant has a slender, woody root. Thereupon Carriere, a French experimenter, sowed the seeds of this wild plant in the autumn, in good soil. The plant found itself in a new predicament. It could not flower before winter came; and, with the elasticity of organization so peculiarly characteristic of natural objects, it formed a thick root, in which was stored nutriment for the growth which must be delayed until the next year. Seeds of these plants, and of their offspring until the fourth generation, were sown, when Carriere found himself in possession of perfect radishes! Now we can picture to ourselves the first radish. Seeds of the wild plant became scattered to a fertile soil. They germinated in the autumn. Some person, more acute than his associates, noticed the sleek, thickened root and tasted it. It pleased him; he watched other plants like it; he sowed the seed. Many biennials,—turnips, carrots, parsnips,—sometimes "break" the first season. Instead of producing fleshy roots, they "run to seed." This appears to be a reversion. Seeds from such plants commonly produce annuals instead of biennials.

The poet Goethe and Saint Hilaire proposed a law which states that when nature expends energy in one direction she spares it in another. There is always an equilibrium of force. There is a constant amount of coin in the treasury; and nature, the scrupulous manager, economizes in stocks when she speculates in crops. We need more proof of this statement; we need to know if it is a law. We are already aware that the number of seeds in a cultivated apple or pear are less than in the wild fruits: do the numbers and sizes of seeds decrease in proportion to increase of improvement? Many fruits have become seedless; man has bred out of the plant the power of perpetuating itself. The banana is a familiar instance. We are familiar with the fact that checking growth induces fruitfulness. We produce fruit at the expense of growth. Old and decrepit apple trees often bear profusely, as if in the endeavor to increase their progeny with their last effort. Poor soil and indifferent culture often produce depauperate plants, and such plants usually blossom prematurely. The intelligent gardener is aware of this fact, and is enabled, in many cases, to produce a race or variety of dwarfs. Here is also a promising field for scientific experiment. Given this law, and we shall sow the smallest seeds, from the fewest-seeded fruits, when we wish to secure new varieties. Here the garden fence was first let down, so far as I know, by the New York Experiment Station.

Running alongside these curious facts are others still more curious. We know that a plant becomes variable when it is cultivated. In their wild condition plants are commonly in a certain state of repose. Grown for centuries under certain conditions, they have become accustomed to their surroundings, fitted into the niche wherein they have found themselves. Their characters become hereditary because they are not disturbed by surrounding objects and conditions. There is a sort of an interbalance between conditions and plants, and when these conditions are changed the probabilities are that the plants will change also. When the tramp got into clean clothes his conditions were changed, and he declined to sleep on the sidewalk. Some plants adapt themselves to new conditions more readily than others; they begin at once to vary in habit and character. Others show no change for some time,—for years, perhaps,—when suddenly they begin to vary, and their original identity may soon be lost. During the first years of their civilization they store up variability which will some day break out into forms whose name is legion. This phenomenon has been called the accumulative effect of cultivation,—a good enough name for an occurrence, a fact, for which we have scarcely a hint of a cause. The direction of variation we can determine largely for ourselves.

Here, for once, does man lead nature; ignorantly, perhaps, but still leads. He leads the plant in the direction of larger roots, sweeter leaves, finer fruits. He could lead more certainly and more rapidly if he knew the whys and wherefores, the bogs and the quicksands, the hard grounds and the mountains, which lie along his path. But with these greater changes come minor ones, which are in some way related to still greater ones which have not appeared. I wish to call your attention to the fact that as variability increases the pollen begins to vary; that delicate, vital dust, which may float in a sunbeam, or which may be carried a thousand miles on the wings of the wind, is influenced by the behavior of the plant. We know little concerning this wonderful fact; we have a hint which is snugly fenced about. It remains for someone to study, develop and apply it. We cannot trace it to its end even in imagination. We do not know whether this is the first or the last of the phenomena of variation. We do not know if every successive generation varies more because the pollen which impregnated, fertilized the seed, varied more. In short, we do not know that this variable pollen does induce variability, although we suppose that pollen from a cultivated plant produces a more variable offspring than does pollen from a wild plant.

But what is this cross-fertilization, this cross-breeding, this hybridizing, which is in every one's mouth, and which flits as an undefined something before the eyes of the farmer of to-day? It is nothing new; its literature is voluminous; gardeners will talk about it in the most commonplace and familiar manner. Its influence has been felt for a quarter of a century as a great tidal wave in the science of horticulture. It is simply transferring the pollen from one flower to another, and then sowing the seeds which result from the fertilization: these seeds will probably produce plants in some manner intermediate between the two parents. This is the gardener's definition. It explains itself; we understand it; we have few or no doubts concerning it. The fresh graduate of the high school has finished natural philosophy; he has learned by rote the definitions and the illustrations in Wells or Quackenbos. He knows it all. In all the book there are no doubts; the statements are definite and positive; there is nothing more to be discovered. Between the covers of our little volume lies all our knowledge of the motions and properties of bodies, of the mechanics of levers and pulleys and wedges and screws, of the wonders of electricity and magnetism, of the laws which govern the weather and the features of the heavens: it is all there. The graduate from the college has studied chemistry and mechanics and physics and electricity and meteorology, but his knowledge is unsatisfactory. He is impressed with what he don't know. His knowledge is relative and negative. He has got beyond the covers of the text-book. He sees every branch of his study widening and widening into infinity: there is no end. We are frittering away our efforts on the surface of this wonderful sexual relation of plant to plant, and are contented if, perchance, we reap a result. The ancient farmer tickled the earth with a sharp stick and was satisfied with his harvest: the farmer before me plows deep; he subsoils; he is never satisfied with his harvest. He succeeds best when he weaves no hit and miss into the acres of his farm. The gardener does not know the laws by which the warp and woof are woven into this mysterious fabric which binds plant to plant in sexual kinship. Said Lindley, a pioneer in horticultural science: "Hybridizing is a game of chance played between man and plants." An Englishman crosses his dahlias and sows the seeds. From 30,000 seedlings, he gets an average of ten good plants! Another calculates that out of 2,000 seedling cross-bred chrysanthemums, he gets, on an average, one good plant. In many cases the ratio of good plants to poor ones is much higher, and in a very few instances we can predict results with tolerable accuracy. Still, the matter is, at best, haphazard; it is not scientific. There are several degrees of crossing, as practiced by man. A transference of the pollen from the anther to the stigma of the same flower is close or self-fertilization; it occurs often in nature, but is rarely practiced by the cultivator. If crossing takes place between different plants of the same species, as between a Baldwin apple and a Swaar apple, or a Marrowfat pea and a Tom Thumb pea, the product of the crossed seeds is styled a half-breed; if it takes place between entirely distinct species, as the pumpkin and the squash, pea and bean, the product is a hybrid.

But now we must ask ourselves what a species is. We must define our definition. The botanist tells us that it is a plant which, in wild nature, reproduces itself, or very nearly itself, from seeds for successive generations. The sugar maple, the apple, the quince, the dahlia are species. But the different sorts of sugar maples,—as the black, the curled, the birdseye,—and the different sorts of apples, quinces, and dahlias are not species: they are varieties; some of more permanent and important character than others. Nature makes species, and also varieties, but man can make only varieties. But we have already seen that man can produce varieties which "come true to seed:" we call them races, but why are they not species? Simply because man has produced them. Many of them we should call species in wild nature. It depends upon which side the garden fence we stand. If we are on the outside, we have a species; if on the inside, we have a race. It is like a Chinese paragraph; if we turn it over, we must stand on our head to read it. The botanist claims the plant when it is a part of wild nature, but loses his interest when it becomes immediately useful to man. Is this a legitimate division of labor? Is the scientist scientific? Does a horse cease to be a horse when it is put into the harness? But they tell us that the different races of cultivated plants—as, for instance, the Treadwell and Clawson wheats, the Yankee and dent corns—are not distinct enough from each other to be called species; and also that, if left to themselves, they will probably soon return into the species from which they sprang. Certainly many of our races are just as distinct from each other as are many reputed wild species, and we have proof that many of them are just as permanent. What do we know of the fixity of wild species, anyway? Scarcely anything. We have many artificial hybrids which, so far as we know, are just as distinct from their parents as their parents are from each other, which are just as fertile, and which appear just as well-fitted to fight out the struggle for existence. We do not know why these hybrids possess such and such characters, characters which are often wholly different from any which appear in the parents. We say that they date back in some mysterious way to ancestors. Then, let us find out what the laws of this reversion are, that we may make other and better species. At present we cross similar species, under apparently identical conditions, but we get different results. Why is it? Is nature fickle, or is man ignorant? Hubbard squashes long grown in Framingham, crossed with Hubbard squashes long grown in Framingham, may improve our seed; but Hubbard squashes long grown in Framingham, crossed with Hubbard squashes newly introduced from Michigan, will infuse new life into our offspring. This crossing with foreign stock of the same variety is of wonderful importance. It is a principle as boundless in its influence as the science of horticulture itself. Its importance may be gleaned from the fact that, in one of Darwin's experiments, the height of foreign crossed stock exceeded that of self-impregnated stock as 100 exceeds 52, and in fertility as 100 exceeds 3. The principle is of universal application, and all honor is due to Darwin who gave it to us. We do not know even the limits at which plants can be crossed.

Sometimes varieties of the same species cannot be crossed, while some species, or reputed species, cross most readily with other species. In short, we know none of the general laws of cross-breeding, and still we believe that there are such laws. If we must learn some of these laws by experiment, we must also learn some from untrained nature. Our woods and fields are nature's garden. For ages the provident mother has been working with winds, and waters, and insects, with soils and climates, to breed up and to breed out her plants. She presents to us a grand puzzle. We do not know whence her plants have come or whither they are tending. We do not know how many are hybrids, born from the beautiful marriage of the insect to the flower; how many are the children of a peculiar clime; how many had their origin in a recent century, or in distant geological time. We are groping, interrogating. Every question which is answered in the woods and fields is answered for the garden. One spirit pervades vegetation. We can scarcely draw a line between cultivation as practiced by man and cultivation as practiced by nature. "Our art," said Shakespeare, "doth mend nature, change it, rather; the art itself is nature." We must get outside the garden fence as well as inside it. We must demolish the line between science and practice. This is the new horticulture. Deep down in nature's heart, beneath the thorns and perplexity, truths are hid which are vital to the farmer and gardener. Then do not discourage the pursuit of science, however much you may have been taught to regard it as opposed to practice. Science is practice. All so-called popular and useful science must be founded upon recondite facts and principles. The more we know of nature as nature, the more readily can we understand nature in the garden.

We fail to catch the butterfly if we chase its irregular flight over the meadow, but the still hunt beside a thistle will bring us a captive. We cannot always reach the result at which we aim in experiment by a direct chase; we quite as often succeed by employing the still hunt of collateral evidence. The experimenter, then, must be a man of skill and learning in more directions than one. To reach the best results he must give his whole time and energies. The college professor, with his classes and his daily routine, can accomplish but little. We must delegate the work to the forthcoming experiment stations.

We commonly look upon the science of botany as affording few avenues for practical research, while we applaud to the skies the results attained by chemistry and entomology. But chemistry often fails just where we expect the greatest results. The chemist finds turnips to be composed largely of water, and declares that they cannot be profitable food for stock; but the old Scotchman, whose turnip-fed sheep are sleek and robust, knows better. The potato is three-fourths water; but it is indispensable, because it presents a digestible bulk to the stomach. Chemistry cannot analyze the grip of a man's stomach. Of all science under heaven, there is none more eminently practical than this same botany. Many people don't know what botany is. They associate it with the school-girl accomplishment, which aims to chase down a few plants to their Latin names, and to press them in a little book, which is sacrilegiously styled an herbarium. This work bears no more relation to botany than does a party platform to party practice. Botany teaches, not only what a plant is, but what it does and how it does it. There is one botany of names and classification, another of cells, another of the plant as a living and growing organism, and another of mutual relations to all environments. All these are given for the use of man, because he deals with plants in all their aspects. Even some botanists tell us that the botany of names and classifications—the botany of species—is well-nigh finished; but when we have named and described every plant upon the face of the earth we must find out what a species is.

The garden is a puzzle. Every leaf and flower is an interrogation point. And why is this true, when we know so many facts in horticulture? Our experiment has been conducted by our so-called practical rather than scientific men. The end and aim of experiment has been to secure more profitable products, rather than to disclose the principles which govern the production of such products. Had we reversed these motives of experiment, had we endeavored to find the why, our horticulture would be much in advance of its present position. Do you understand me? Do you understand that it is more necessary, at present, to discover laws than to strive directly for better fruits and vegetables?

The difficulties in horticulture keep pace with the advancements in horticulture; the more we know the more we do not know. We shall experiment and investigate for a century; we shall solve the riddles of to-day: what, then, shall the horticulturist of the future investigate? We do not know what his puzzles will be, but we know that he will have puzzles. Science is ever new. It has no depth, no height, no boundaries; it stretches away into the infinite. We no sooner uncover one truth than we discover another. Man always anticipates his extremity of want but never reaches it. Before we exhaust the coal and oil which mother earth has locked in her bosom, we grasp the electric current from the air. Before we shall exhaust our iron and copper we shall learn an easy method of extracting the silver from clay. Man shall always strive. Endeavor is a winsome goddess, who leads us through copses and along hazardous banks, but she never leads us to the ends of nature. The man who loves his garden, and who knows some of its secrets, is impatient for a fuller gratification. Some objects are near at hand and well defined, others are misty on the horizon. He tries to grasp them; they flit away like a pleasant dream; the prosaic garden fence is before him.

Appendix II
Books by Liberty Hyde Bailey

Titles are listed chronologically by year, and then alphabetically when published in the same year. Bailey was a strong believer in writing and editing books in series, so, when applicable, the series title and any meaningful relationship to other books in the list are indicated. The number of books Bailey wrote vacillates depending on how you count; this list attempts to account for all books with Bailey listed as a primary author or editor, including meaningful rewrites under new titles. We have identified seventy-six such works, three of which are multivolume cyclopedias. Many of these books went through numerous revisions and sometimes significant expansions that are not accounted for here. It should also be kept in mind that Bailey is estimated to have penned some 1,300 articles published in periodicals ranging from popular farm magazines to the journal *Science*, compiled over 100 additional papers of pure taxonomy, edited at least 117 titles by ninety-nine different authors (including books in The Rural Science Series, The Rural Manuals, The Rural Text-Book Series, The Rural State and Province Series, and The Open-Country Books), and founded and edited a number of significant periodicals, including *Country Life in America* and *Gentes Herbarum*.[1]

The symbol † indicates books cited in this volume.

† *Talks Afield: About Plants and the Science of Plants.* Boston: Houghton, Mifflin, 1885.

Field Notes on Apple-Culture. New York: Orange Judd, 1886.

† *The Garden Fence.* Boston: Wright and Potter, 1886.

The Horticulturist's Rule-Book: A Compendium of Useful Information for Fruit-Growers, Truck-Gardeners, Florists, and Others. New York: Garden Publishing, 1889. In 1892 a section reprinted as a separate pamphlet, *Injurious Insects and Plant Diseases, with Remedies.* Revised in 1895 for inclusion in The Garden-Craft Series. Revised and radically expanded in 1911 to form *The Farm and Garden Rule-Book* for inclusion in The Rural Manuals.

Annals of Horticulture in North America for the Year 1889: A Witness of Passing Events and a Record of Progress. Annals of Horticulture 1. New York: Rural Publishing, 1890.

Annals of Horticulture in North America for the Year 1890. Annals of Horticulture 2. New York: Rural Publishing, 1891.

The Nursery-Book: A Complete Guide to the Multiplication and Pollination of Plants. New York: Rural Publishing, 1891. Revised in 1896 for inclusion in The Garden-Craft Series (under the title *The Nursery-Book: A Complete Guide to the Multiplication of Plants*). By 1907 also included in The Rural Science Series.

Annals of Horticulture in North America for the Year 1891. Annals of Horticulture 3. New York: Rural Publishing, 1892.

Cross-Breeding and Hybridizing: The Philosophy of the Crossing of Plants, Considered with Reference to Their Improvement under Cultivation; with a Brief Bibliography of the Subject. The Rural Library. New York: Rural Publishing, 1892. In 1895 combined with several lectures to form the first edition of *Plant-Breeding.*

American Grape Training; An Account of the Leading Forms Now in Use of Training the American Grape. New York: Rural Publishing, 1893.

Annals of Horticulture in North America for the Year 1892. Annals of Horticulture 4. New York: Rural Publishing, 1893.

Annals of Horticulture in North America for the Year 1893: Comprising an Account of the Horticulture of the Columbian Exposition. Annals of Horticulture 5. New York: Orange Judd, 1894.

Plant-Breeding: Being Six Lectures upon the Amelioration of Domestic Plants. The Garden-Craft Series. New York: Macmillan, 1895. Included and expanded upon the 1892 text of *Cross-Breeding and Hybridizing.* Editions from 1915 on, revised by Arthur W. Gilbert and without the original sub-title, included in The Rural Science Series.

The Survival of the Unlike: A Collection of Evolution Essays Suggested by the Study of Domestic Plants. New York: Macmillan, 1896. Companion to *Sketches of the Evolution of Our Native Fruits* (1898).

The Forcing-Book: A Manual of the Cultivation of Vegetables in Glass Houses. The Garden-Craft Series. New York: Macmillan, 1897. By 1914 also included in The Rural Science Series.

Lessons with Plants: Suggestions for Seeing and Interpreting Some of the Common Forms of Vegetation. New York: Macmillan, 1897.

The Principles of Fruit-Growing. The Rural Science Series. New York: Macmillan, 1897.

First Lessons with Plants: Being an Abridgement of "Lessons with Plants." New York: Macmillan, 1898.

† *Garden-Making: Suggestions for the Utilizing of Home Grounds.* The Garden-Craft Series. New York: Macmillan, 1898. Revised and combined with *The Practical Garden-Book* in 1910 to form *Manual of Gardening* as part of The Rural Manuals.

The Principles of Agriculture: A Text-Book for Schools and Rural Societies. Edited by L. H. Bailey. The Rural Science Series. New York: Macmillan, 1898.

The Pruning-Book: A Monograph of the Pruning and Training of Plants as Applied to American Conditions. The Garden-Craft Series. New York: Macmillan, 1898. By 1911 included in The Rural Science Series. Revised in 1916 as *The Pruning-Manual* as part of The Rural Manuals.

Sketch of the Evolution of Our Native Fruits. New York: Macmillan, 1898. Companion to *The Survival of the Unlike* (1896).

Botany: An Elementary Text for Schools. New York: Macmillan, 1900. Companion to *Botany for Secondary Schools* (1910).

† *The Gardener: A Book of Brief Directions for the Growing of the Common Fruits, Vegetables and Flowers in the Garden and about the Home.* New York: Macmillan, 1900. Succeeded in 1934 by *The Gardener's Handbook.*

The Practical Garden-Book: Containing the Simplest Directions for the Growing of the Commonest Things about the House and Garden. By Charles E. Hunn and L. H. Bailey. The Garden-Craft Series. New York: Macmillan, 1900. Revised and combined with *Garden-Making* in 1910 to form *Manual of Gardening* under Bailey's name as part of The Rural Manuals.

† *The Principles of Vegetable-Gardening.* The Rural Science Series. New York: Macmillan, 1901.

† *Cyclopedia of American Horticulture: Comprising Suggestions for Cultivation of Horticultural Plants, Descriptions of the Species of Fruits, Vegetables,*

Flowers and Ornamental Plants Sold in the United States and Canada, Together with Geographical and Biographical Sketches. Edited by L. H. Bailey. 4 vols. New York: Macmillan, 1900–1902. Succeeded in 1914–17 by *The Standard Cyclopedia of Horticulture.*

Nature-Portraits: Studies with Pen and Camera of Our Wild Birds, Animals, Fish and Insects. New York: Doubleday, Page, 1902. Featured "Text by the Editor of *Country-Life in America*" (with Bailey's name nowhere listed) as well as many visual artists who had contributed to that magazine under Bailey's editorship. Text revised in 1903 to form Part II of *The Nature-Study Idea.*

† *The Nature-Study Idea: Being an Interpretation of the New School-Movement to Put the Child in Sympathy with Nature.* New York: Doubleday, Page, 1903. Part II consisted of the text of *Nature-Portraits* (1902). Revised in 1909 (with a new subtitle) for The Rural Outlook Set 2.

† *The Outlook to Nature.* New York: Macmillan, 1905. Revised in 1911 for The Rural Outlook Set 1.

Beginners' Botany. New York: Macmillan, 1908.

First Course in Biology. By L. H. Bailey and W. M. Coleman. New York: Macmillan, 1908.

Poems. New York: The Cornell Countryman, 1908.

The State and the Farmer. The Rural Outlook Set 3. New York: Macmillan, 1908.

Cyclopedia of American Agriculture: A Popular Survey of Agricultural Conditions, Practices and Ideals in the United States and Canada. Edited by L. H. Bailey. 4 vols. New York: Macmillan, 1907–9. Vols. II and III reprinted separately with new introductions in 1922 as *Cyclopedia of Farm Crops: A Popular Survey of Crops and Crop-Making Methods in the United States and Canada* and *Cyclopedia of Farm Animals,* respectively.

The Training of Farmers. New York: Century, 1909.

Botany for Secondary Schools: A Guide to the Knowledge of the Vegetation of the Neighborhood. New York: Macmillan, 1910. Companion to *Botany* (1900).

† *Manual of Gardening: A Practical Guide to the Making of Home Grounds and the Growing of Flowers, Fruits, and Vegetables for Home Use.* The Rural Manuals. New York: Macmillan, 1910. A "combination and revision of the main parts of" *Garden-Making* (1898) and *The Practical Garden-Book* (1900), "together with much new material and the results of the experience of ten added years."

The Country-Life Movement in the United States. The Rural Outlook Set 4. New York: Macmillan, 1911.

Farm and Forest. Edited by L. H. Bailey. Vocations Series. Boston: Hall & Locke, 1911.

Farm and Garden Rule-Book: A Manual of Ready Rules and Reference, with Recipes, Precepts, Formulas, and Tabular Information for the Use of General Farmers, Gardeners, Fruit-Growers, Stockmen, Dairymen, Poultry-Men, Foresters, Rural Teachers, and Others in the United States and Canada. The Rural Manuals. New York: Macmillan, 1911. A revision and radical expansion of *The Horticulturist's Rule-Book* (1889).

Outlook. Ithaca, NY: Self-published, 1911. [Verse.]

Report of the Commission on Country Life. By Commission on Country Life (L. H. Bailey, Chairman; Henry Wallace; Kenyon L. Butterfield; Gifford Pinchot; Walter H. Page; Charles S. Barrett; and William A. Beard). New York: Sturgis & Walton, 1911.

York State Rural Problems I. The Problem Books 1. Albany: J. B. Lyon, 1913.

† *The Standard Cyclopedia of Horticulture: A Discussion, for the Amateur, and the Professional and Commercial Grower, of the Kinds, Characteristics and Methods of Cultivation of the Species of Plants Grown in the Region of the United States and Canada for Ornament, for Fancy, for Fruit and for Vegetables; with Keys to the Natural Families and Genera, Descriptions of the Horticultural Capabilities of the States and Provinces and Dependent Islands, and Sketches of Eminent Horticulturists.* Edited by L. H. Bailey. 6 vols. New York: Macmillan, 1914–17. Based on the 1900–1902 *Cyclopedia of American Horticulture,* this was "a new work with an enlarged scope." Condensed in 1925 to three volumes in a "New Edition."

† *The Holy Earth.* The Background Books: The Philosophy of the Holy Earth 1. New York: Charles Scribner's Sons, 1915.

York State Rural Problems II. The Problem Books 2. Albany: J. B. Lyon, 1915.

Ground-Levels in Democracy. Ithaca, NY: Self-published, 1916.

† *The Pruning-Manual: Being the Eighteenth Edition, Revised and Reset, of The Pruning-Book, Which Was First Published in 1898.* The Rural Manuals. New York: Macmillan, 1916. A revision of *The Pruning-Book* (1898).

† *Wind and Weather.* The Background Books: The Philosophy of the Holy Earth 2. New York: Charles Scribner's Sons, 1916. [Verse.] A handful of these poems republished in 1952 as *My Great Oak Tree and Other Poems.*

† *Home Grounds: Their Planning and Planting.* Harrisburg: J. Horace McFarland, 1918.

RUS: Rural Uplook Service, a Preliminary Attempt to Register the Rural Leadership in the United States and Canada. Vol. 1. Ithaca, NY: Self-published, 1918.

† *Universal Service: The Hope of Humanity.* Ithaca, NY: Self-published; Comstock Publishing, Agents, 1918. By 1919 subtitle removed and book included in The Background Books: The Philosophy of the Holy Earth 3.

What Is Democracy? The Background Books: The Philosophy of the Holy Earth 4. Ithaca, NY: Self-published; Comstock Publishing, Agents, 1918.

The Nursery-Manual: A Complete Guide to the Multiplication of Plants. The Rural Manuals. New York: Macmillan, 1920. A revision and significant expansion of *The Nursery-Book* (1891).

RUS: A Register of the Rural Leadership in the United States and Canada. Vol. 2. Ithaca, NY: Self-published, 1920.

† *The School-Book of Farming: A Text for the Elementary Schools, Homes, and Clubs.* The Rural Text-Book Series. New York: Macmillan, 1920.

† *The Apple-Tree.* The Open Country Books 1. New York: Macmillan, 1922.

The Cultivated Evergreens: A Handbook of the Coniferous and Most Important Broad-Leafed Evergreens Planted for Ornament in the United States and Canada. New York: Macmillan, 1923. Succeeded in 1933 by *The Cultivated Conifers.*

The Seven Stars. The Background Books: The Philosophy of the Holy Earth 5. New York: Macmillan, 1923.

Manual of Cultivated Plants: A Flora for the Identification of the Most Common or Significant Species of Plants Grown in the Continental United States and Canada for Food, Ornament, Utility, and General Interest, Both in the Open and under Glass. The Rural Manuals. New York: Macmillan, 1924.

RUS: A Biographical Register of Rural Leadership in the United States and Canada. Vol. 3. By L. H. Bailey and Ethel Zoe Bailey. Ithaca, NY: Self-published, 1925.

† *The Harvest of the Year to the Tiller of the Soil.* The Background Books: The Philosophy of the Holy Earth 6. New York: Macmillan, 1927.

† *The Garden Lover.* The Background Books: The Philosophy of the Holy Earth 7. New York: Macmillan, 1928.

Hortus: A Concise Dictionary of Gardening, General Horticulture and Cultivated Plants in North America. Edited by L. H. Bailey and Ethel Zoe Bailey. New York: Macmillan, 1930.

RUS: A Biographical Register of Rural Leadership in the United States and Canada. Vol. 4. Edited by L. H. Bailey and Ethel Zoe Bailey. Ithaca, NY: Self-published, 1930.

The Cultivated Conifers in North America: Comprising the Pine Family and the Taxads. New York: Macmillan, 1933. "Successor to" *The Cultivated Evergreens* (1923).

How Plants Get Their Names. New York: Macmillan, 1933.

Gardener's Handbook: Brief Indications for the Growing of Common Flowers, Vegetables and Fruits in the Garden and about the Home. New York: Macmillan, 1934. "Successor to" *The Gardener* (1900).

Supplement to Hortus, for the Five Current Years Including 1930. Edited by L. H. Bailey and Ethel Zoe Bailey. New York: Macmillan, 1935.

† *The Garden of Gourds, with Decorations.* The Garden Books 1. New York: Macmillan, 1937.

† *The Garden of Pinks, with Decorations.* The Garden Books 2. New York: Macmillan, 1938.

The Garden of Larkspurs, with Decorations. The Garden Books 3. New York: Macmillan, 1939.

† *Hortus Second: A Concise Dictionary of Gardening, General Horticulture and Cultivated Plants in North America.* Edited by L. H. Bailey and Ethel Zoe Bailey. New York: Macmillan, 1941.

My Great Oak Tree and Other Poems. Waltham, MA: Chronica Botanica, 1952. Selections from *Wind and Weather* (1916).

The Garden of Bellflowers, with Decorations. The Garden Books 4. New York: Macmillan, 1953.

Notes

Preface

1. In *Words Said about a Birthday: Addresses in Recognition of the Ninetieth Anniversary of the Natal Day of Liberty Hyde Bailey, Delivered at Cornell University, April 29, 1948*, p. 26, MS 21-2-3342, Liberty Hyde Bailey Papers, 1845–2004, Division of Rare and Manuscript Collections, Cornell University Library.

2. L. H. Bailey, *The Nature-Study Idea: Being an Interpretation of the New School-Movement to Put the Child in Sympathy with Nature* (New York: Doubleday, 1903), 29.

3. Andrew Denny Rodgers III, *Liberty Hyde Bailey: A Story of American Plant Sciences* (1949) (New York: Hafner, 1965), 86.

4. Bailey, *Nature-Study Idea*, 5.

5. I treat the significance of Bailey's use of the term "biocentric" in my notes to the centennial edition of *The Holy Earth* (Berkeley, CA: Counterpoint, 2015), 113n22. Bailey describes his work as "earth-philosophy" in *Universal Service* (1918) (Ithaca, NY: Self-published, 1919), 16.

6. I describe Bailey's significance for current environmental and humanistic scholarship in more detail in my introduction and endnotes to the centennial edition of *The Holy Earth*.

Introduction

1. Rodgers, *Liberty Hyde Bailey*, 97. The text of "The Garden Fence" is provided in appendix 1 in this volume.

2. Rodgers, 96.

3. Quoted without citation in Philip Dorf, *Liberty Hyde Bailey: An Informal Biography* (Ithaca, NY: Cornell University Press, 1956), 82.

4. These quotations are scattered variously among the following books: Bailey, *Holy Earth*, 4; Bailey, *The Harvest of the Year to the Tiller of the Soil* (New York: Macmillan, 1927), 119; and Bailey's essay "Blossoms" included in this anthology.

5. Russell Lord and Kate Lord, *Forever the Land: A Country Chronicle and Anthology* (New York: Harper, 1950), 108–9.

II. The Growing of the Plants

1. "Candytuft" is the chapter following "307" in Bailey, *The Garden Lover* (New York: Macmillan, 1928).

2. Now commonly spelled *Linanthus*.

3. Here and elsewhere, we follow Bailey's preferred spelling, capitalization, and italicization. According to the *Chicago Manual of Style*, a more modern presentation would capitalize and italicize singular genus names, italicize a genus and species combination with the genus as a capitalized initial and the species in all lowercase, and leave plural genus names both roman and in all lowercase. Bailey's preference for roman over italic lettering in all but genus and species combinations may reflect his desire that everyday gardeners learn Latinate plant names and adopt them into English without fear, rather than think of them as part of a foreign (and perhaps intimidating) language.

4. Now commonly spelled *Pycnostachys*.

III. Flowers

1. A photograph of four-o'clocks originally accompanied this essay.

2. Now commonly spelled "cobaeas."

IV. Fruits & Vegetables

1. The "inventory" Bailey mentions is a nine-page vegetable garden plant list grouped by herbage, root, and fruit vegetables.

2. [Bailey's note] See also the recent extensive volume issued by the N.Y. Agric. Exper. Station (Geneva), called "Sturtevant's Notes on Edible Plants."

3. "The Elm Tree," *Country Life in America*, January 1903, 120.

4. Lawton B. Evans, Luther N. Duncan, and George W. Duncan, eds., *Farm Life Readers* (Boston: Silver, Burdett, 1913), book 5.

V. Spring to Winter

1. More commonly known today as the codling moth.

2. Bailey, *The Garden of Gourds* (New York: Macmillan, 1937).

3. Bailey, *The Garden Lover* (New York: Macmillan, 1928).

VI. Epilogue

1. In his later correspondence with the founders of the Friends of the Land, Bailey renamed his proposal "Society of the Whole Earth" (Lord and Lord, *Forever the Land*, 108–9; and see the introduction to this volume, pp. 9–10).

Appendix I

1. Attributed to *The Gardener's Chronicle*, in A. A. Crozier, *The Modification of Plants by Climate* (Ann Arbor: Self-published, 1885), 24.

Appendix II

1. These numbers are taken from Harlan P. Banks, "Liberty Hyde Bailey: 1858–1954," *Biographical Memoirs of the National Academy of Sciences* 64 (1994): 16. Banks's essay also includes a partial publication list.

Bibliography

I. The Garden in the Mind

Epigraph. In *The Gardener: A Book of Brief Directions for the Growing of the Common Fruits, Vegetables and Flowers in the Garden and about the House*, v. New York: Macmillan, 1925.

"General Advice." In *Garden-Making: Suggestions for the Utilizing of Home Grounds*, 1–7. The Garden-Craft Series. New York: Macmillan, 1898.

"To One Who Hath No Garden." In *The Garden Lover*, 20–22. The Background Books: The Philosophy of the Holy Earth 7. New York: Macmillan, 1928.

"The Common Natural History." In *The Outlook to Nature*, 37–39. New and rev. ed. The Rural Outlook Set 1. New York: Doubleday, 1911.

"The Importance of Seeing Correctly." In *Talks Afield: About Plants and the Science of Plants*, 152–54. Boston: Houghton, Mifflin, 1885.

"A Reverie of Gardens." *The Outlook*, June 1, 1901, 267–76.

"The Feeling for Plants." In *Home Grounds: Their Planning and Planting*, 14–16. Harrisburg: J. Horace McFarland, 1918.

"Planting a Plant." In *Nature-Study Quarterly*, 367–68. Ithaca, NY: College of Agriculture, Cornell University, 1901.

"Gardening and Its Future." 1930. MS 21-2-3342, Liberty Hyde Bailey Papers, 1845–2004, Division of Rare and Manuscript Collections, Cornell University Library.

"Undertone." In *Wind and Weather*, 212. The Background Books: The Philosophy of the Holy Earth 2. New York: Charles Scribner's Sons, 1916.

II. The Growing of the Plants

Epigraph. In *The Gardener: A Book of Brief Directions for the Growing of the Common Fruits, Vegetables and Flowers in the Garden and about the House*, v. New York: Macmillan, 1925.

"The Miracle." In *The School-Book of Farming: A Text for the Elementary Schools, Homes, and Clubs,* 91–92. The Rural Text-Book Series. New York: Macmillan, 1920.

"How to Make a Garden—The First Lesson." *Country Life in America,* November 1901, 31–32.

"The Home Garden." Introduction to *How to Make a Flower Garden,* edited by Wilhelm Miller, ix-xxii. New York: Doubleday, 1903. The original title of this piece was "The Spirit of the Home Garden." It has been changed in this volume to avoid confusion with "The Spirit of the Garden."

"How to Make a Garden—Digging in the Dirt." *Country Life in America,* April 1902, 217–18.

"The Growing of Plants by Children—The School Garden." In *The Nature-Study Idea,* 78–84. 4th rev. ed. The Rural Outlook Set 2. New York: Macmillan, 1913.

"307." In *The Garden Lover,* 33–44. The Background Books: The Philosophy of the Holy Earth 7. New York: Macmillan, 1928.

"The Spirit of the Garden." In *The Outlook to Nature,* 76–79. New and rev. ed. The Rural Outlook Set 1. New York: Macmillan, 1911.

"Oak." In *Cyclopedia of American Horticulture,* 3:1110. New York: Macmillan, 1901.

"The Principles of Pruning." In *The Pruning-Manual,* 107–8. The Rural Manuals. New York: Macmillan, 1936.

"The Weather." In *The Outlook to Nature,* 33–37. New rev. ed. The Rural Outlook Set 1. New York: Macmillan, 1911.

"What Is a Weed?" 1947. MS 21-2-3342, Liberty Hyde Bailey Papers, 1845–2004, Division of Rare and Manuscript Collections, Cornell University Library.

"White Clover." In *Wind and Weather,* 70. The Background Books: The Philosophy of the Holy Earth 2. New York: Charles Scribner's Sons, 1916.

III. Flowers

Epigraph. "A Talk about Dahlias." Address, Cornell University, Ithaca, New York, February 22, 1897. In *Annual Report of the Department of Agriculture,* 5:99. 2nd ed. New York: State Printer, 1898.

"Blossoms." In *The Garden Lover,* 25–27. The Background Books: The Philosophy of the Holy Earth 7. New York: Macmillan, 1928.

"The Symbolism of Flowers." *American Garden,* January 1891, 170.

"Extrinsic and Intrinsic Views of Nature." In *The Nature-Study Idea,* 124–30. 4th rev. ed. The Rural Outlook Set 2. New York: Macmillan, 1913.

"The Flower-Growing Should Be Part of the Design." In *Manual of Gardening,* 27–36. New York: Macmillan, 1910. Essay previously published in *The*

China Asters, with Remarks upon Flower-Beds (Ithaca, NY: Cornell University, 1895).

"Annuals: The Best Kinds and How to Grow Them." In *How to Make a Flower Garden*, edited by Wilhelm Miller, 3–15. New York: Doubleday, 1903.

"Campanula." In *Wind and Weather*, 64–65. The Background Books: The Philosophy of the Holy Earth 2. New York: Charles Scribner's Sons, 1916.

IV. Fruits & Vegetables

Epigraph. "Markets and Catalogs." In *The Garden Lover*, 110. The Background Books: The Philosophy of the Holy Earth 7. New York: Macmillan, 1925.

"The Admiration of Good Materials." In *The Holy Earth*, 69–75. Centennial ed. Berkeley, CA: Counterpoint, 2015. Originally published in 1915 by C. Scribner's Sons, New York.

"The Affection for the Work." In *The Principles of Vegetable-Gardening*, 12–14. The Rural Science Series. New York: Macmillan, 1921.

"The Growing of the Vegetable Plants." In *Manual of Gardening: A Practical Guide to the Making of Home Grounds and the Growing of Flowers, Fruits, and Vegetables for Home Use*, 451–54. The Rural Manuals. New York: Macmillan, 1916.

"The Fruit-Garden." In *The Garden Lover*, 82–99. The Background Books: The Philosophy of the Holy Earth 7. New York: Macmillan, 1928.

"Peach." In *The Harvest of the Year to the Tiller of the Soil*, 165–68. The Background Books: The Philosophy of the Holy Earth 6. New York: Macmillan, 1927.

"Where There Is No Apple-Tree." In *The Apple-Tree*, 7–9. The Open Country Books 1. New York: Macmillan, 1922.

"Apple-Year." In *Wind and Weather*, 71–73. The Background Books: The Philosophy of the Holy Earth 2. New York: Charles Scribner's Sons, 1916.

V. Spring to Winter

Epigraph. "The Goals." In *The Garden Lover*, 154. The Background Books: The Philosophy of the Holy Earth 7. New York: Macmillan, 1928.

"The Garden Flows." *House and Garden*, March 1944, 23, 89.

"The New Year." In *The Harvest of the Year to the Tiller of the Soil*, 3–6. The Background Books: The Philosophy of the Holy Earth 6. New York: Macmillan, 1927.

"The Dandelion." In *First Lessons with Plants: Being an Abridgement of "Lessons with Plants,"* 68–74. New York: Macmillan, 1898.

"The Apple Tree in the Landscape." In *The Apple-Tree*, 10–14. The Open Country Books 1. New York: Macmillan, 1922.

"Lessons of To-day." Address, Anniversary Field Day, Geneva, New York, August 29, 1907. In *Annual Report of the Board of Control of the New York Agricultural Experiment Station*, 3:52. 26th ed. Albany: J. B. Lyon, 1908.

"Leaves." In *The Harvest of the Year to the Tiller of the Soil*, 157–58. The Background Books: The Philosophy of the Holy Earth 6. New York: Macmillan, 1927.

"The Garden of Gourds." In *The Garden of Gourds, with Decorations*, 1–10. The Garden Books 1. New York: Macmillan, 1937.

"Lesson I—The Pumpkin." In *Junior Naturalist Monthly*, 1–2. Ithaca, NY: College of Agriculture, Cornell University, 1904.

"November: June." In *The Harvest of the Year to the Tiller of the Soil*, 124–26. The Background Books: The Philosophy of the Holy Earth 6. New York: Macmillan, 1927.

"An Outlook on Winter." In *The Nature-Study Idea*, 161–67. 4th rev. ed. The Rural Outlook Set 2. New York: Macmillan, 1913. Essay previously published in *Country Life in America*, December 1901, 37–40.

"Midwinter." In *The Nature-Study Review*, 373. Ithaca, NY: Comstock Publishing, 1916.

"Greenhouse in the Snow." In *The Garden Lover*, 69–75. The Background Books: The Philosophy of the Holy Earth 7. New York: Macmillan, 1928.

"The Garden of Pinks." In *The Garden of Pinks, with Decorations*, 1–6. The Garden Books 2. New York: Macmillan, 1938.

"December." In *Wind and Weather*, 176–78. The Background Books: The Philosophy of the Holy Earth 2. New York: Charles Scribner's Sons, 1916.

Epilogue

"Marvels at Our Feet." In *The Land: A Quarterly Magazine*, Spring 1945, 145–49.

"Society of the Holy Earth." In *Universal Service*, 164–65. The Background Books: The Philosophy of the Holy Earth 3. Ithaca, NY: Comstock Publishing, 1919.

Appendix I

The Garden Fence. Boston: Wright & Potter, 1886.

Index

Page numbers in *italics* indicate illustrations and their captions. Works without citation are Bailey's.